Gravity, Black Holes, and the Very Early Universe

T0202841

Tai L. Chow
Author

Gravity, Black Holes, and the Very Early Universe

An Introduction to General Relativity and Cosmology

Springer

Tai L. Chow
California State University Stanislaus
Turlock, CA
USA

ISBN-13: 978-1-4419-2525-1   e-ISBN-13: 978-0-387-73631-0

Dedicated to My Wife
Sylvia Hsiu Yung Chen

About the Author

Dr. Tai Chow was born and raised in China. He received a Bachelor of Science degree in physics from National Taiwan University, a Master's degree in physics from Case Western Reserve University in Cleveland, and a Ph.D. in physics from the University of Rochester in New York. Since 1970, Dr. Chow has been in the Department of Physics at California State University, Stanislaus, and served as department chairman for 17 years. He has also served as a Visiting Professor at the University of California at Davis and Berkeley and has worked as a Summer Faculty Fellow at Stanford University and at NASA/Ames Center. Dr. Chow has published over 40 articles in physics and astrophysics journals and is the author of three textbooks: *Classical Mechanics,* published in 1995 by John Wiley & Sons, *Mathematical Methods for Physicists,* in 2000 by Cambridge University Press, and *Introduction to Electromagnetic Theory,* in 2005 by Jones & Bartlett Publishers.

Preface

In the early 1900s, three events took place that dramatically changed the course of modern physics. In 1905 Albert Einstein formulated the Special Theory of Relativity. Then, in 1915, he developed the General Theory of Relativity, and around 1925 quantum mechanics took its present form. Since then, physics has progressed rapidly. Beginning in 1930, quantum mechanics and special relativity were united into what is known as the relativistic quantum field theory. This merger was very rewarding in that it provides, at the least, partial explanation of the laws and interactions governing elementary particle physics.

Among the four types of forces (strong, electromagnetic, weak, and gravitational) known today, gravity is perhaps the strangest. Weak though it is, gravity dominates the other three forces over cosmic distances. Any cosmology must be founded on a logically secure theory of gravitation.

The first three forces could be explained through particle interactions taking place in the flat space-time of special relativity. However, gravity defies such an explanation. In order to describe the mysterious force known as gravity, Einstein in 1915 was compelled to generalize the ideas of his special relativity, and he eventually connected gravity with the geometry of space-time. In other words, Einstein's General Theory of Relativity is a relativistic theory of gravitation.

For a long time, Einstein's Theory of General Relativity occupied an isolated position within the domain of general physics. This was attributable in part to the mathematical framework of the theory, which is based on Riemannian geometry, a kind of geometry not needed in most other physical applications. The extreme difficulty in devising suitable experiments that might verify the theory and the growth of more fertile fields of investigation, such as atomic and nuclear physics as well as the study of elementary particles, also contributed to the isolation of the theory.

However, Einstein's Theory of General Relativity is now enjoying renewed interest. This is due partly to the development of new technological capabilities that opened up previously inaccessible avenues for the experimental verification of general relativity and partly to the conjecture of some theoretical physicists that the fundamental difficulties confronting quantum field theory may find their resolution in a suitable combination of the two disciplines. The discovery of extremely compact celestial objects—neutron stars and black holes, for instance—provided the final turning point. The study of these objects demanded the application of Einstein's Theory of General Relativity. Today, physics and astronomy have joined forces to form the discipline called relativistic astrophysics. Einstein's Theory of General Relativity is also essential to modern cosmology, since the overall space-time structure is intimately related to the gravitational field. In the past decade interest in cosmology and general relativity has grown considerably.

Today, there is increased demand for undergraduate courses in relativity and cosmology. There are many advanced books on the Theory of General Relativity and cosmology for the specialist, and many elementary expositions for the lay reader. But there is a gap at the undergraduate level. This book is an attempt to fill the gap. We will try to make available to the student a working acquaintance with the concepts and fundamental ideas in general relativity and modern cosmology. For the modes of calculation we choose the old-fashioned tensor calculus for pedagogical reasons. Most undergraduates have not been exposed to the many new formalisms developed in general relativity. Hopefully after reading this book, the student can continue delving more deeply into particular aspects or topics in general relativity and cosmology that interest him or her.

This book evolved from a set of lecture notes for a course that I have taught over the past 10 years. I am making the assumption that the student has been exposed to a calculus-based course in general physics and a course in calculus (including the handling of differentiations of field equations). Some exposure to tensor analysis would be helpful but is not necessary; this subject is covered in the text.

The student will find that in the derivations of equations, a generous amount of detail has been given. However, to ensure that the student does not lose sight of the development underway, some of the more lengthy and tedious algebraic manipulations have been omitted.

Turlock, California   *Tai L. Chow, Ph.D.*

Contents

Chapter 1
Basic Ideas of General Relativity

1.1 Inadequacy of Special Relativity

Einstein's special relativity rejected the ether concept of a privileged inertial frame of reference, which still depended on the concept of inertial frames. What is so special about these frames? This remained a mystery in special relativity, and it was reserved for general relativity to solve, or at least to elucidate, this problem.

In establishing his general relativity, Einstein was influenced by Ernst Mach. Mach's ideas on absolute space and inertia are roughly these: (a) space is simply the separation between bodies, and time is merely the succession of events, so neither space nor time has an independent existence in its own right and relative motions are all that matter; (b) the property of inertia has nothing to do with absolute space but arises from some kind of interaction (unspecified by Mach) between each individual body and all the other matter in the universe. If there were no other masses, an isolated body would have no inertia. This contrasted with Newton's view that the body would still have inertia because of the effect of absolute space. Einstein was very impressed by this whole complex of ideas expressed by Mach and coined the term "Mach's principle" to describe them.

The inertia concept dates back to Galileo, but it is expressed formally by Newton's laws of motion. An object will continue to be in a state of rest or of uniform motion unless acted upon by some external force. The acceleration that is caused by a given force is inversely proportional to the mass of the object:

$$\vec{F} = m\vec{a}. \tag{1.1}$$

Here $\vec{F}$ is the applied force, $\vec{a}$ the acceleration produced, and m the mass of the body. The more "inert" the body, the greater will be the force required to change its state of rest or of motion.

To measure the velocity and acceleration of the object we need a frame of reference within which we can note down the displacement of the object at successive times. If (1.1) is valid in frame S_0, is it also valid in another frame, S_1, that has an acceleration $\vec{A}$ relative to S_0? Relative to S_1 we have

$$\vec{F}_1 = m\vec{a}_1 \tag{1.2}$$

but we also have

$$\vec{a} = \vec{a}_1 + \vec{A} \tag{1.3}$$

and so (1.2) becomes

$$\vec{F}_1 = \vec{F} - m\vec{A}. \tag{1.4}$$

Since the extra term depends solely on the mass of the object, Newton called it the inertial force, and the frame of reference S_0 the inertial frame. All other frames that are not accelerated relative to S_0 are also inertial frames. Frames like S_1 that require inertial force corrections are non-inertial frames.

In practice we merely specify an approximate inertial frame in accordance with the needs of the problem under investigation. For elementary applications in the lab, a frame attached to Earth usually suffices. This frame is an approximate inertial frame, owing to the daily rotation of Earth on its axis and its revolution around the sun.

There are two ways of measuring Earth's rotation about its polar axis. We can measure this rotation by setting up a Foucault pendulum, whose plane gradually rotates around a vertical axis as the pendulum swings, with an angular velocity Ω

$$\Omega = \omega \sin \lambda \tag{1.5}$$

where λ is the latitude of the place of observation and ω the angular velocity of Earth's rotation. Knowing λ and Ω, ω can be determined.

We can also measure Earth's rotation about its axis relative to distant stars. By observing the rising and setting of stars astronomers can determine the rotation period of Earth around its axis. The two methods give the same angular velocity. At first sight this doesn't seem surprising. But closer examination reveals why the result is nontrivial. The second method measures Earth's rotation period against a background of distant stars. The first method, the Foucault pendulum, employs the standard Newtonian mechanics in a rotating frame of reference and takes note of how Newton's laws of motion get modified. Thus, implicit in the assumption that equates the two methods is that to give a consistent picture of Newton's laws of motion, we need the background of the distant parts of the universe. Mach attached great significance to this observation in the last century.

Mach argued that the concept of inertia has status solely because of the background provided by the universe. If there were no background, we could not identify S_0 in preference to other frames. Consider a single object in an otherwise empty universe. In the absence of any force Newton's law becomes

$$m\vec{a} = 0. \tag{1.6}$$

If we conclude from this that $\vec{a} = 0$, then the object is moving with uniform motion relative to S_0. But we now no longer have a background against which to measure velocities. Mach argued that the correct conclusion to be drawn from (1.6) is if $m = 0$, then $\vec{a}$ is indeterminate.

That is, the measure of inertia is somehow related to the background produced by the universe. Remove the background, and the inertia disappears! So inertia is not just a property of matter, as is usually assumed in Newtonian mechanics. By relating the inertia of matter to the distant parts of the universe, Mach destroyed the purely local character of the laboratory.

This whole complex of ideas expressed by Mach impressed Einstein very much, who coined the term "Mach's principle" to describe them, and they provided a fruitful source of conjectures for Einstein to investigate quantitatively. But he soon realized that, through the work of de Sitter, general relativity theory and Mach's principle are not fully compatible, so he seems to have abandoned Mach's principle.

Let us go back to the concept of inertial frames of reference, which obviously is incompatible with gravitational phenomena. An inertial frame is defined as one in which a free particle (i.e., no force acting on it) moves with a constant velocity. But gravity is long-range and cannot be screened. Consequently, the only way to visualize an inertial frame is to imagine it far away from any matter. A concept like this is clearly of little use to someone doing experiments on Earth or to astronomers whose observations relate to distant massive galaxies. Attempts to modify Newtonian gravity to make it compatible with special relativity have not been successful. According to Einstein, if gravitation is long-range and unscreened, it has something of a permanent character, and it must be intrinsic to the region in which it is located. Einstein identified this intrinsic property of space-time by its geometry.

We know that the geometry of space-time is pseudo-Euclidean in special relativity. In the presence of matter the space-time geometry, according to Einstein, should be non-Euclidean. Therefore, he sought to relate the intrinsic parameters of non-Euclidean space-time geometry to the distribution of matter and energy. Thus, in general relativity the gravitational effects will not be described through an explicit external force but rather through the non-Euclidean nature of the space-time geometry.

General relativity is unquestionably a more modern theory of gravitation, and it has supplanted Newton's. Since both special relativity and Newtonian gravitation represent an approximation of the truth, we expect that general relativity should reduce to special relativity in cases where the gravitational effects are small, and should resemble Newtonian gravitation when special relativistic effects are negligible.

1.2 Einstein's Principle of Equivalence

Although we cannot introduce inertial frames of reference in the strict Newtonian sense, in real situations we can cover space-time with a patchwork of local inertial frames. The introduction of local inertial frames depends on the equivalence of inertial and gravitational mass. The mass entering in Newton's second law is referred to as initial mass m_I, while the mass appearing in the law of gravity is called the gravitation mass. When gravity acts on a body, it acts on the gravitational mass m_g, and the result of the force is an acceleration of the inertial mass m_I. The fact that all bodies fall in a vacuum with the same acceleration indicates that within experimental accuracy, the ratio of inertial to gravitational mass is independent of the body. Newton realized this even when he formulated his laws of motion, and he was able to show that, if there was a difference between gravitational and inertial mass, it was not greater than one part in 1,000. If the fractional amount by which inertial and gravitational masses can differ is indicated by x, then Newton showed that $x < 1/1,000$. Bessel observed that the periods T of pendulums made of different materials of lengths l offered a sensitive test of the equivalence:

$$T = 2\pi \sqrt{\frac{m_I}{m_g} \frac{l}{g}}.$$

By measuring the periods of pendulums of equal lengths at the same place (the gravitational acceleration g is the same), he showed that $x < 1/(6 \times 10^4)$. Eötvos used a very ingenious method to set the limit more precisely. If a pendulum is set at a latitude λ as shown in Fig. 1.1, the direction of the pendulum will not be toward the center of Earth but in the direction of the resultant of the forces $m_I \omega^2 R$ and $m_g g$. If θ is the angle between the direction of the pendulum and the direction to the center of Earth, then θ is a simple function of m_I/m_g. Eötvos used a null method to compare the ratio m_I/m_g for different objects. His equipment consisted of a torsion pendulum (Fig. 1.2). By orienting the axis $m_1 m_2$ to the north, south, east, and west successively, and observing the twist in the torsion fiber with a mirror and microscope, any inequality of m_{I1}/m_{I2} for equal values of m_{g1}/m_{g2} may be determined. Eötvos established that x is less than $(1/2) \times 10^{-8}$. Dicke redid the Eötvos experiment with modern equipment and reported that $x < 1/10^{10}$. Braginski and Panov of Russia reported that $x < 1/10^{12}$.

How can the equivalence of the inertial and gravitational mass of the same body make it possible to establish the local inertial frames? To see how, let us consider a region of space-time in which a constant gravitational field $\vec{g}$ exists. One such region is the neighborhood of any point on the surface of Earth. If gravity were the only

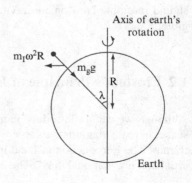

Fig. 1.1 A pendulum at a latitude of λ.

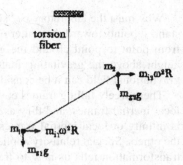

Fig. 1.2 Eötvos's torsion pendulum.

force acting, all bodies in the region would fall with the same acceleration $\vec{a} = \vec{g}$. Thus, by transforming to a frame f^γ with acceleration $\vec{g}$ we can eliminate the effects of gravitation; any object will appear unaccelerated unless a nongravitational force $\vec{F}_{ng}$ acts on it. It is easy to show this formally. Consider a particle inside a stationary elevator cabin acted upon by a gravitational force $m\vec{g}$ and a nongravitational $\vec{F}_{ng}$; Newton's law gives

$$m\vec{a} = m\vec{g} + \vec{F}_{ng}.$$

If the elevator cabin is severed from its supporting cable and allowed to fall freely down a long shaft under the action of Earth's gravity, in frame f^γ (the accelerated cabin) the acceleration of the particle relative to the cabin is $\vec{a}' = \vec{a} - \vec{g}$, and so

$$m(\vec{a}' + \vec{g}) = m\vec{g} + \vec{F}_{ng}$$

or

$$m\vec{a}' = \vec{F}_{ng} \tag{1.7}$$

In (1.7), gravitational forces do not enter, i.e., gravity has been "transformed away" in the free-fall cabin (frame f^γ). Therefore, the equivalence of gravitational and inertial mass implies the following:

> *In a small laboratory falling freely in a gravitational field, mechanical phenomena are the same as those observed in an inertial frame in the absence of a gravitational field.*

In 1907 Einstein generalized this conclusion by replacing the word "mechanical phenomena" with "the laws of physics," and the resulting statement is known as the *principle of equivalence*.

Einstein's conclusion may at first seem difficult to accept. The principle seems to be derived from the Newtonian expression that we know is not rigorously valid. We do not derive it at all. We are only inclined toward it by the approximately valid Newtonian theory. Further, as the conclusion actually depends only on the equivalence of gravitational and inertial mass, it can be valid even when we reject the Newtonian mechanics. Einstein's equivalence principle, as above, is identical in content to the equivalence of inertial and gravitational mass. They are two different ways of formulating the same principle. Einstein's equivalence principle is sometimes called the "strong" equivalence principle, and the equivalence of inertial and gravitational mass is known as the "weak" equivalence principle.

Why must the laboratory be "small?" Because real gravitation fields are not constant: $\vec{g}$ points toward the center of the gravitating matter, and so its direction varies from point to point around the gravitating matter, and its magnitude varies with height above the gravitating matter. Only over a sufficiently small region of the gravitational field can it be considered uniform.

These freely falling frames covering the neighborhood of an event are called the local inertial frames, and they are very important in relativity. Near an event there is infinity of local inertial frames, each moving with constant velocity relative to the others. Special relativity applies rigorously within these frames, and the Lorentz transformation tells us how to transform the coordinates of an event from one of these frames to another.

It is obvious that local inertial frames are not inertial in the strict Newtonian sense, because (1.7) only corresponds to Newton's second law if the gravitational force $m\vec{g}$ is ignored. They are more restricted and also more general than Newton's inertial frames: more restricted, because the in-homogeneity of real gravitational fields makes them only locally applicable, instead of infinite in extent; more general, because any laboratory in free fall (for instance Skylab) is a local inertial frame—it does not have to be unaccelerated relative to the galaxies or to absolute space.

It is very important to bear in mind that we can only "transform away" gravitation in a limited region of space-time by employing a laboratory in free-fall. This cannot be accomplished on a large scale. No laboratory exists that can cover all the space around a gravitating object and move in such a way as to eliminate its effects.

Can we create the effects of gravity by choosing a suitable frame of reference? The answer is a definite yes! Consider an observer inside a closed cabin. When this cabin is fixed on Earth's surface, the observer will feel his or her normal weight, and will observe that all falling bodies accelerate toward the floor at the same rate (Fig. 1.3a). Now if this observer were placed in an identical closed cabin out in space far from all massive bodies (so that no gravitational field exists), and if the cabin were fitted with a rocket motor capable of accelerating it smoothly at a rate exactly equal to the gravitational acceleration on Earth, he or she would again find that all free objects accelerated toward the floor at the same rate and would also feel his or her normal weight (Fig. 1.3b). There are no observations or experiments the observer could perform in the cabin that could indicate whether the effects were

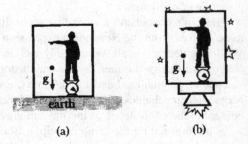

Fig. 1.3 Equivalence of gravitation and acceleration.   (a)   (b)

those of gravity or those of acceleration. Within a small closed cabin the effects of gravity and acceleration are indistinguishable.

Gravity in this context behaves like an inertial force, a fictitious force that arises due to the acceleration of the frame of reference from which observations are being made. The most familiar examples of inertial forces are the centrifugal and Coriolis forces in a rotating system fixed on Earth's surface.

1.3 Immediate Consequences of the Principle of Equivalence

The principle of equivalence leads to two testable conclusions about the propagation of light. If the effects of gravitation and acceleration are indistinguishable, then rays of light should bend in a gravitational field. As well, light moving up through a gravitational field should be redshifted. Let us look at these phenomena in turn.

1.3.1 The Bending of a Light Beam

Consider an observer inside an enclosed space cabin that is accelerating through a gravitation-free region of space. A light ray enters the cabin from a window at P, and it is parallel to the floor of the cabin as it enters. Where will the light hit on the opposite wall? The cabin will move upward while the light is traveling to the opposite wall so that the light will hit the wall at Q', a little below Q. Thus, to an observer inside the cabin, the light ray curves downward as it travels through the cabin (Fig. 1.4a). Now using the equivalence principle, we conclude that in a gravitational field, light will not travel along a straight line, but its path will be curved (Fig. 4b). Near Earth's surface, where g is locally constant and parallel, and where this fall of light is far too small to be measurable; for instance, a horizontally projected beam, after traveling 1 km, has fallen only about 1 Å. It is possible to detect the deflection of light falling past the sun, but this involves the patching together of many local inertial frames, and we put off this calculation for future work.

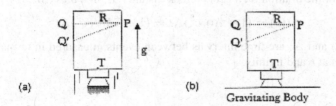

Fig. 1.4 Equivalence principle predicts fall of light.

1.3.2 Gravitational Shift of Spectral Lines (Gravitational Redshift)

Consider again the space cabin, traveling in a gravitation-free region of space with the acceleration $\vec{g}$. T and R are two fixed points on a straight line parallel to the direction of $\vec{g}$. A light wave of frequency ν is emitted at T. This light wave will not have the same frequency when it reaches R. To analyze the situation we first note that the light ray will take a time $\Delta t = h/c$ to reach R (h is the distance between T and R), and during this time Point R gains an additional velocity $\Delta v = g\Delta t$. This apparent relative velocity with respect to T will cause a frequency shift $\Delta \nu$:

$$\frac{\Delta \nu}{\nu} = \frac{\Delta v}{c} = \frac{gh}{c^2}. \tag{1.8}$$

Now, using the equivalence principle, we can conclude that the same phenomenon must be observed in a gravitational field. We note that gh is the difference in gravitational potential between source and receiver. Let us denote this difference by $\Delta \phi$. We see that in passing up through a gravitational potential difference of $\Delta \phi$ the light has become redder and its wavelength is increased by the amount $\Delta \phi/c^2$. If light falls through a gravitational potential difference $\Delta \phi$, it gains energy and becomes bluer by the amount $\Delta \phi/c^2$, i.e., its wavelength is Doppler-shifted toward the blue by the amount $\Delta \phi/c^2$.

Pound and Rebka verified this remarkable prediction in a terrestrial laboratory in 1960. They allowed a 14.4 keV γ-ray, emitted by the radioactive decay of ^{57}Fe, to fall 22.6 m down an evacuated tower, and measured the change in its frequency. The predicted blueshift is $z = -2.46 \times 10^{-15}$, and they measured $z = (-2.57 \pm 0.26) \times 10^{-15}$, thus directly verifying the equivalence principle. Such high precision was possible because of the Mössbauer effect. This is the emission of radiation from an atomic nucleus in a crystal, which gives a spectral line with a very precisely defined frequency.

The gravitational shift of spectral lines implies that in a gravitational field a clock (a periodic phenomenon) runs slower than does the same clock in a gravitation-free region of space. We can regard a radiating atom as a clock, each "tick" being the emission of a wave crest. Since light from a clock in a gravitational field will be reddened when received by "clock at infinity," the latter clock will see the former clock ticking more slowly than itself if the clocks are of identical construction. The gravitational time dilation factor for a clock distant r from a mass M is

$$\Delta t(r) = (1 - \Delta \phi/c^2)\Delta t = (1 - GM/rc^2)\Delta t \tag{1.9}$$

where $\Delta t(r)$ and Δt are time intervals between events, measured in terms of ticks of the clocks at r and infinity.

1.4 The Curved Space-Time Concept

Special relativity has familiarized us with space-time in terms of geometry, and the space-time of special relativity is described by the Minkowskian metric

$$ds^2 = c^2 d\tau^2 = c^2 dt^2 - (dx^2 + dy^2 + dz^2)$$

where $d\tau$ is a measure of the proper time. Now, since $t(r)$ of (1.9) represents the time read by a clock at rest at r, this must be the proper time; that is, for a clock at rest

$$ds^2 = c^2 d\tau^2 = \left(1 - 2GM/rc^2\right)c^2 dt^2 \tag{1.10}$$

apart from higher orders. Since the intervals of proper time are affected by gravitational field, we may say that the presence of a gravitational field influences the geometry of space-time, and to the extent to which (1.4) is valid, we may write

$$ds^2 = (1 - 2GM/rc^2)c^2 dt^2 - (dx^2 + dy^2 + dz^2) \tag{1.11}$$

and we see that the presence of the gravitational field modifies the structure of space-time, which is no longer Minkowskian. In fact, the proof of the existence of the gravitational redshift rules out the possibility of a theory of gravitation in Minkowski space. As soon as we assume that the gravitational field influences time, we must also assume the possibility of influencing the measure of its length. Space and time coordinates must be treated on equal footing without any intrinsic preference of one over the other. (In fact, we expect the theory to reduce to Minkowski space in the limit of the gravitational field.) Thus, we may expect the general line element to be of the form

$$ds^2 = Ac^2 dt^2 - (Bdx^2 + Cdy^2 + Ddz^2) \tag{1.12}$$

where the coefficients A, B, C, and D depend on the gravitation and are functions of the space-time variables x^0, x^1, x^2, x^3; they reduce to unity at large distances from the gravitating source. But a gravitational field is equivalent to a certain non-inertial frame of reference, and when we use a non-inertial frame, the four-dimensional coordinate system is curvilinear. For example, if (x', y', z', ct') are the coordinates of a frame rotating uniformly with angular velocity ω along the z-axis, the transformation to this frame is given by

$$x = x' \cos \omega t - y' \sin \omega t, \ y = x' \sin \omega t + y' \cos \omega t, \ z = z'$$

and the line element ds^2 becomes

$$ds^2 = [c^2 - \omega^2(x'^2 + y'^2)]dt^2 - dx'^2 - dy'^2 - dz'^2 + 2\omega y' dx' dt - 2\omega x' dy' dt.$$

It is evident that this expression cannot be represented as a sum of squares of the coordinate differentials. Therefore, in a non-inertial frame of reference, the line-element is, in general, a quadratic form in the differentials of the coordinates x^0, x^1, x^2, and x^3 of the general type:

$$ds^2 = \sum_{\mu,\nu} g_{\mu\nu} dx^\mu dx^\nu = g_{\mu\nu} dx^\mu dx^\nu \tag{1.13}$$

where the $g_{\mu\nu}$ are functions of the space-time variables x^0, x^1, x^2, x^3, and are called the metric of the space-time manifold. The geometry of a space-time in which a metric form such as (1.13) can be defined is called a *Riemannian geometry*.

As accelerated (non-inertial) frames are equivalent to gravitational fields, gravitational effects are to be described by the metric $g_{\mu\nu}$. In this framework, gravitation is to be understood as a deviation of the metric of the space-time manifold from the flat Minkowski metric. Therefore, the metric $g_{\mu\nu}$ is not fixed arbitrarily on the whole space-time, but, as we will see, it depends on the local distribution of matter.

It is clear that the metric $g_{\mu\nu}$ is symmetric in the indexes μ and ν($g_{\mu\nu} = g_{\nu\mu}$). In the general case, there are 10 different quantities $g_{\mu\nu}$ - 4 with equal indexes, 6 with different indexes. In inertial frames, when we use Cartesian coordinates, the quantities $g_{\mu\nu}$ are locally

$$g_{00} = 1, g_{11} = g_{22} = g_{33} = -1, g_{ik} = 0 \text{ for } i \neq k \qquad (1.14)$$

The choice of sign for the metrics is not standard; others may use $g_{00} = -1$ and $g_{ii} = 1$ for $I = 1, 2, 3$. We call a four-dimensional system of coordinates with these values of $g_{\mu\nu}$ Galilean. By an appropriate choice of coordinates, we can always bring the metrics $g_{\mu\nu}$ to Galilean form at any point of the non-Galilean space-time. We note that, after reduction to diagonal form at a given point, the matrix of the quantities $g_{\mu\nu}$ has one positive and three negative principal values. This set of signs is called the signature of the matrix. The determinant g, formed from the quantities $g_{\mu\nu}$, is always negative for a real space-time: g < 0.

We wish to stress a fundamental difference between "real" gravitational fields and non-inertial frames, in spite of their local equivalence. A real gravitational field cannot be eliminated by any coordinate transformation. In other words, in the presence of a gravitational field, space-time is such that the quantities $g_{\mu\nu}$ cannot, by any coordinate transformation, be brought to their Galilean values over all space-time. Such a space-time is said to be curved.

Let us here explain the idea of curvature. We have enough difficulty in imagining a curved three-dimensional space, let alone a curved four-dimensional space-time, so let us start with two-dimensional surfaces, with which we are familiar. Obviously, a plane is flat. Consider a sphere or a cylinder: are these surfaces flat or curved? The sphere cannot be deformed to coincide with a plane without stretching or tearing, so a sphere is fundamentally curved. The curvature of the cylinder is less fundamental, because it can be simply unrolled onto a plane without distortion.

In these three cases, we reached our conclusion by considering the three surfaces as embedded in three-dimensional space. It was the great achievement of Gauss in the early 19th century to discover that the curvature and the whole geometry of a surface could be determined by employing the metric tensor $g_{\mu\nu}$. In a plane, the distance between two points separated by dx^1, dx^2 is

$$ds^2 = g_{\mu\nu}dx^\mu dx^\nu = (dx^1)^2 + (dx^2)^2 \qquad (1.15)$$

where

$$g = \begin{pmatrix} 1 & 0 \\ 0 & 1 \end{pmatrix}. \qquad (1.15a)$$

Pythagoras's theorem is satisfied, indicating that we are in a flat space. If we were given the metric distance in the polar coordinates r and θ

$$ds^2 = dr^2 + r^2 d\theta^2$$

then the metric tensor is position-dependent

$$g = \begin{pmatrix} 1 & 0 \\ 0 & r^2 \end{pmatrix}.$$

How can we tell whether the surface is flat or not? By employing the following coordinate transformation

$$x^1 = r\cos\theta, \text{ and } x^2 = r\sin\theta$$

we would get back the Cartesian distance and metric tensor formulas (1.15) and (1.15a). Now consider a cylinder of radius R. With cylindrical coordinates r, z, ϕ in three-dimensional embedding space (Fig. 1.5), the surface is defined by $r = R = $ constant, and then

$$ds^2 = dz^2 + R^2 d\phi^2.$$

Now, if we define coordinates x^1 and x^2:

$$x^1 = z, x^2 = R\phi, \text{ then } ds^2 = (dx^1)^2 + (dx^2)^2,$$

which is the same as the distance formula on a plane, the surface of a cylinder is therefore intrinsically flat.

Next let us look at the surface of a sphere of radius R. We can use spherical coordinates θ and ϕ to describe positions on the surface of the sphere:

$$x^1 = \theta, x^2 = \phi$$

Then we have

$$ds^2 = R^2 d\theta^2 + R^2 \sin^2\theta d\phi^2 = R^2(dx^1)^2 + R^2 \sin^2\theta(dx^2)^2$$

and the metric

$$g = \begin{pmatrix} R^2 & 0 \\ 0 & R^2 \sin x^1 \end{pmatrix}.$$

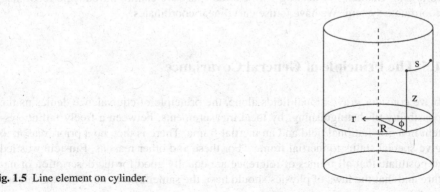

Fig. 1.5 Line element on cylinder.

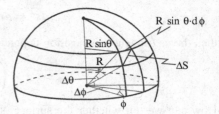

Fig. 1.6 Line element on sphere.

Now, we test for flatness by seeking new coordinates x'^1, x'^2, which are functions of x^1 and x^2 (i.e., of θ and ϕ); in terms of these new coordinates the distance takes the Cartesian form

$$(dx'^1)^2 + (dx'^2)^2.$$

But however hard we try, we cannot find such a coordinate transformation, and it appears that the metrical properties of a spherical surface are intrinsically different from those of a plane. How can we know that there are no obscure transformations that will reduce a spherical surface to a plane? Can we know the curvature from the metric tensor alone? Gauss gave the answer. He first showed that for the two-dimensional curved space the metric tensor can always be transformed to diagonal form, where $g_{12} = g_{21} = 0$, and the metric is called *orthogonal*. He then showed that the curvature K of the surface is given by the formula

$$K = \frac{1}{2g_{11}g_{22}} \left\{ -\frac{\partial^2 g_{11}}{\partial (x^2)^2} - \frac{\partial^2 g_{22}}{\partial (x^1)^2} + \frac{1}{2g_{11}} \left[\frac{\partial g_{11}}{\partial x^1} \frac{\partial g_{22}}{\partial x^1} + \left(\frac{\partial g_{11}}{\partial x^2} \right)^2 \right] \right.$$
$$\left. + \frac{1}{2g_{22}} \left[\frac{\partial g_{11}}{\partial x^2} \frac{\partial g_{22}}{\partial x^2} + \left(\frac{\partial g_{22}}{\partial x^1} \right)^2 \right] \right\} \tag{1.16}$$

Obviously K is zero for a plane, either with Cartesian coordinates or polar coordinates, and $K = 1/R^2$ for the spherical surface of radius R. However, in spaces of more than two dimensions it is not possible to specify curvature by only one function K. It turns out that a fourth-rank curvature tensor $R_{\nu\lambda\beta}$ is necessary. It is defined in terms of derivatives of the metric tensor and the derivation is given in the following chapter. For dealing with curved space we cannot introduce a rectilinear coordinate system. We have to use curvilinear coordinates.

1.5 The Principle of General Covariance

If we consider gravitational fields alone, the principle of equivalence denies us the possibility of distinguishing, by local measurements, between a freely falling system in a gravitational field and an inertial frame. There is then no a priori reason to give special status to inertial frames. For these and other reasons, Einstein was led to postulate that all frames of reference are equally good for the description of nature and that the laws of physics should have the same form in all. The equivalence of all frames of reference must be represented by the equivalence of all coordinate systems, since reference frames are represented by coordinate systems. The requirement that, under a general coordinate transformation, the laws of physics must remain covariant (i.e., form invariant) is called the *principle of general covariance*. It is the mathematical representation of the principle of equivalence. If we use tensorial quantities for expressing the laws of physics, the principle of general covariance will yield the "simplest" tensor equations that generalize the special relativistic versions.

As will be demonstrated in the next chapter, if we have a tensor, say $T^{\mu\nu}$, defined in the coordinate system $S(x^0, x^1, x^2, x^3)$, the tensor will transform into

$$T'^{\mu\nu} = \frac{\partial x'^\mu}{\partial x^\alpha} \frac{\partial x'^\nu}{\partial x^\beta} T^{\alpha\beta}$$

under a coordinate transformation from S to $S'(x'^0, x'^1, x'^2, x'^3)$. Since this is a completely linear form in the components of the tensor $T^{\mu\nu}$, the vanishing of all of its components in one coordinate system leads to the vanishing of all of its components in any other coordinate system. Therefore, let us consider, say in S, the tensor equation that represents a physics law,

$$A^{\mu\nu} = B^{\mu\nu}.$$

We now write

$$C^{\mu\nu} = A^{\mu\nu} - B^{\mu\nu}$$

And then, transforming this into coordinate system S', we have

$$C'^{\mu\nu} = \frac{\partial x'^\mu}{\partial x^\alpha} \frac{\partial x'^\nu}{\partial x^\beta} C^{\alpha\beta}.$$

As a consequence of the tensor equation, the vanishing of $C^{\mu\nu}$ in the unprimed coordinate system S leads to the vanishing of $C'^{\mu\nu}$ in the primed system

$$C'^{\mu\nu} = A'^{\mu\nu} - B'^{\mu\nu} = 0$$

or

$$A'^{\mu\nu} = B'^{\mu\nu}$$

That is, the physics law remains form invariant.

1.6 Distance and Time Intervals

In the Theory of General Relativity there exists no restriction of any kind regarding the choice of a permissible coordinate system. So the question naturally arises that given a certain coordinate system $x^\mu (\mu = 0, 1, 2, 3)$, how do we relate the actual time and distance to these coordinates? Let us first find the relation of the proper

time, denoted by τ as before, to the coordinate x^0. In this respect let us consider two
infinitesimally separated events. If the two events take place at one and the same
point in space, then the interval $ds (= g_{\mu\nu}dx^\mu dx^\nu)$ between the two events is $cd\tau$,
where $d\tau$ is the proper time interval between the two events. Stated differently, the
proper time interval is the interval of time as measured by the clock of an observer
who is at rest (at a point in space) in the gravitational field. Setting $dx^1 = dx^2 =
dx^3 = 0$ in the general expression $ds = g_{\mu\nu}dx^\mu dx^\nu$, we get the connection between
$d\tau$ (the element of proper time) and dx^0 (the coordinate differential):

$$d\tau = \frac{1}{c}\sqrt{g_{00}}dx^0 \tag{1.17}$$

or for the time between any two events occurring at the same point in space

$$\tau = \frac{1}{c}\int \sqrt{g_{00}}dx^0. \tag{1.18}$$

This relation determines the actual time interval (i.e., the proper time for the given
point in space) for a change of the coordinate x^0.

We next determine the element d$ of spatial distance. In the Special Theory of
Relativity we can define dl as the interval between two infinitesimally separated
events occurring at one and the same time. However, we cannot define dl in this
way in the General Theory of Relativity, because in a gravitational field the proper
time at different points in space has a different dependence on the coordinate x^0. To
find dl, we can proceed as follows:

Consider two infinitesimally close points in space, A and B; A has coordinates x^μ
and B has coordinates $x^\mu + dx^\mu$. An observer at B sends a light signal to A and then
receives it back over the same path in space (Fig. 1.7). The time (as measured by the
observer at B) required for this, when multiplied by c, is the distance between A and
B. The interval ds between two events corresponding to the departure and arrival of
a light signal from one to the other is equal to zero:

$$ds^2 = g_{\mu\nu}dx^\mu dx^\nu = 0 \tag{1.19}$$

or

$$g_{00}(dx^0)^2 + 2g_{0i}dx^i dx^0 + g_{ij}dx^i dx^j = 0, i, j = 1, 2, 3 \tag{1.19a}$$

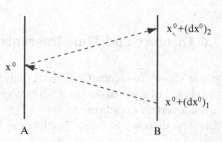

Fig. 1.7 World lines of observers A and B, and
the signals.

where we have separated the space and time coordinates. We find two roots for the co-ordinate time dx^0:

$$(dx^0)_1 = \frac{1}{g_{00}} \left[-g_{0i} dx^i - \sqrt{(g_{0i}g_{0j} - g_{ij}g_{00})dx^i dx^j} \right]. \tag{1.20}$$

$$(dx^0)_2 = \frac{1}{g_{00}} \left[-g_{0i} dx^i + \sqrt{(g_{0i}g_{0j} - g_{ij}g_{00})dx^i dx^j} \right] \tag{1.21}$$

If x^0 is the moment of arrival of the signal at A, the times when it left B and when it will return to B are, respectively, $x^0 + (dx^0)_1$ and $x^0 + (dx^0)_2$. In Fig. 1.7 the solid lines are the world lines corresponding to the given coordinates x^μ and $x^\mu + dx^\mu$, and the dashed lines are the world lines of the signals. Therefore the total lapse of the coordinate time for the journey back and forth is equal to

$$(dx^0)_2 - (dx^0)_1 = \frac{2}{g_{00}} \sqrt{(g_{0i}g_{0j} - g_{ij}g_{00})dx^i dx^j} \tag{1.22}$$

and the corresponding interval of proper time is

$$\frac{1}{c}\sqrt{g_{00}} \left[(dx^2)_2 - (dx^0)_1 \right]. \tag{1.23}$$

We thus obtain, for the spatial distance dl between two infinitesimally separated points, the expression

$$dl^2 = \left(-g_{ij} + g_{0i}g_{0j}/g_{00} \right) dx^i dx^j. \tag{1.24}$$

We rewrite it in the form

$$dl^2 = \gamma_{ij} dx^i dx^j. \tag{1.25}$$

with

$$\gamma_{ij} = \left(-g_{ij} + g_{0i}g_{0j}/g_{00} \right) \tag{1.26}$$

The metric tensor $g_{\mu\nu}$ generally depends on x^0, so the space metric dl^2 also changes with time. For this reason, it is meaningless to integrate dl, as such an integration would depend on the world line chosen between the two given space points. Thus, generally speaking, in the General Theory of Relativity the concept of a definite distance between two bodies loses its meaning, remaining valid only for infinitesimal separations. Only when the metric tensor $g_{\mu\nu}$ does not depend on the time (and the distance can also then be defined over a finite domain) can the integral dl along a space curve have a definite meaning.

1.7 Problems

1.1. By direct transformation from Cartesian coordinates, calculate the metric tensor $g_{\mu\nu}$ in a flat three-dimensional space in terms of (a) spherical coordinates and (b) cylindrical coordinates.

1.2. Consider the redshift produced upon light escaping from the gravitational field of a celestial body. Show that the fractional change in wavelength of the light that escapes from the celestial body is $\Delta\lambda/\lambda = GM/Rc^2$, where G = universal gravitational constant, R = radius of the celestial body, and M = mass.

1.3. Using Gauss's curvature formula Eq. (10), show that $K = 0$ for a plane when plane polar coordinates are used, and that $K = 1/R^2$ for a sphere of radius a when spherical coordinates are used.

1.4. Calculate the curvature $K(r)$ of a surface whose metric distance formula is

$$\Delta s^2 = f(r)\Delta r^2 + r^2\Delta\phi^2.$$

1.5. Use the principle of equivalence to answer the following two questions:

(a) Suppose you have just lighted a candle in an elevator when, unfortunately, the cable breaks. The elevator falls freely. What happens to the candle flame?
(b) As shown in Fig. 1.8, a brass ball attached to a spring hangs outside a metal cup into which the ball can fit snugly. The spring passes through a hole in the cup and down through a pipe, where it is tied to a spring. This entire assembly is mounted on a curtain rod so that one can hold on to the whole contraption easily. Finally the cup and ball assembly is enclosed in a transparent glass sphere. By design, the spring is too weak to counteract the force of gravity, and so the ball hangs limply outside the cup. Find a surefire way to pop the ball into the cup every time.

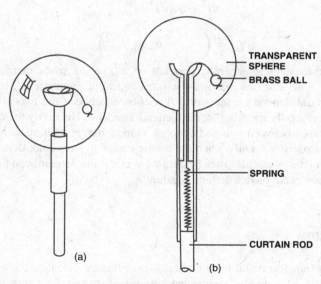

Fig. 1.8 Einstein's toy.

References

Berry M (1976) *Principles of Cosmology and Gravitation* (Cambridge University Press, Cambridge, England)

Landau LD, Lifshitz EM (1975) *The Classical Theory of Fields.* (Pergamon Press Ltd., Oxford, England)

Chapter 2
Curvilinear Coordinates and General Tensors

2.1 Curvilinear Coordinates

We devote this chapter to the development of four-dimensional geometry in arbitrary curvilinear coordinates. We shall deal with field quantities. A field quantity has the same nature at all points of space. Such a quantity will be disturbed by the curvature.

If we take a point quantity Q (or one of its components if it has several), we can differentiate with respect to any of the four coordinates. We write the result

$$\frac{\partial Q}{\partial x^\mu} = Q_{,\mu}.$$

A subscript preceded by a comma will always denote a derivative in this way. We put the index downstairs in order that we may maintain a balancing of the indexes in the general equations. We can see this balancing by noting that the change in Q, when we move from the point x^μ to a neighboring point $x^\mu + dx^\mu$, is

$$\delta Q = Q_{,\mu} \delta x^\mu \tag{2.1}$$

where summation over μ is understood. The repeated indexes appearing once in the lower and once in the upper position are automatically summed over; this is Einstein's summation convention.

Let us consider the transformation from one coordinate system x^0, x^1, x^2, x^3 to another x'^0, x'^1, x'^2, x'^3:

$$x^\mu = f^\mu(x'^0, x'^1, x'^2, x'^3) \tag{2.2}$$

where the f^μ are certain functions. When we transform the coordinates, their differentials transform according to the relation

$$dx^\mu = \frac{\partial x^\mu}{\partial x'^\nu} dx'^\nu. \tag{2.3}$$

Any set of four quantities $A^\mu (\mu = 0, 1, 2, 3)$ which, under coordinate change, transform like the coordinate differentials, is called a *contravariant vector:*

$$A^\mu = \frac{\partial x^\mu}{\partial x'^\nu} A'^\nu. \tag{2.4}$$

If ϕ is somewhat scalar, under a coordinate change, the four quantities $\partial \phi / \partial x^\mu$ transform according to the formula

$$\frac{\partial \phi}{\partial x^\mu} = \frac{\partial x'^\nu}{\partial x^\mu} \frac{\partial \phi}{\partial x'^\nu}. \tag{2.5}$$

Any set of four quantities $A_\mu (\mu = 0, 1, 2, 3)$ that, under a coordinate transformation, transform like the derivatives of a scalar is called a *covariant vector*:

$$A_\mu = \frac{\partial x'^\nu}{\partial x^\mu} A'_\nu \tag{2.6}$$

From the two contravariant vectors A^μ and B^μ we may form the 16 quantities $A^\mu B^\nu (\mu, \nu = 0, 1, 2, 3)$. These 16 quantities form the components of a contravariant tensor of the second rank: Any aggregate of 16 quantities $T^{\mu\nu}$ that, under a coordinate transformation, transform like the products of two contravariant vectors

$$T^{\mu\nu} = \frac{\partial x^\mu}{\partial x'^\alpha} \frac{\partial x^\nu}{\partial x'^\beta} T'^{\alpha\beta} \tag{2.7}$$

is a contravariant tensor of rank two. We may also form a covariant tensor of rank two from two covariant vectors, which transform according to the formula

$$T_{\mu\nu} = \frac{\partial x'^\alpha}{\partial x^\mu} \frac{\partial x'^\beta}{\partial x^\nu} T'_{\alpha\beta}. \tag{2.8}$$

Similarly, we can form a mixed tensor T_ν^μ of order two that transforms as follows

$$T_\nu^\mu = \frac{\partial x^\mu}{\partial x'^\alpha} \frac{\partial x'^\beta}{\partial x^\nu} T'^\alpha_\beta. \tag{2.9}$$

We may continue this process and multiply more than two vectors together, taking care that their indexes are all different. In this way we can construct tensors of higher rank. The total number of free indexes of a tensor is called its *rank* (or *order*).

We may set a subscript equal to a superscript and sum over all values of this index, which results in a tensor having two fewer free indexes than the original one. This process is called *contraction*. For example, if we start with a fourth-order tensor $T^\mu{}_{\nu\rho}{}^\sigma$, one way of contracting it is to put $\sigma = \rho$, which gives the second-rank tensor $T^\mu{}_{\nu\rho}{}^\rho$, having only 16 components, arising from the four values of μ and ν. We could contract again to get the scalar $T^\mu{}_{\mu\rho}{}^\rho$ with just one component.

It is easy to show that the inner product of contravariant and covariant vectors, $A_\mu B^\mu$, is an invariant, that is, independent of the coordinate system

$$A'_\mu B'^\mu = \frac{\partial x^\alpha}{\partial x'^\mu} \frac{\partial x'^\mu}{\partial x^\beta} A_\alpha B^\beta = \frac{\partial x^\alpha}{\partial x^\beta} A_\alpha B^\beta = \delta_\beta^\alpha A_\alpha B^\beta = A_\alpha B^\alpha.$$

The square of the line element in curvilinear coordinates is a quadratic form in the differentials dx^μ : $ds^2 = g_{\mu\nu}dx^\mu dx^\nu$. Since the contracted product of $g_{\mu\nu}$ and the contravariant tensor $dx^\mu dx^\nu$ is a scalar, the $g_{\mu\nu}$ forms a covariant tensor:

$$ds^2 = g'_{\alpha\beta}dx'^\alpha dx'^\beta = g_{\alpha\beta}dx^\alpha dx^\beta$$

Now, $dx'^\alpha = (\partial x'^\alpha/(\partial x^\mu)dx^\mu$, so that

$$g'_{\alpha\beta}\frac{\partial x'^\alpha}{\partial x^\mu}\frac{\partial x'^\beta}{\partial x^\nu}dx^\mu dx^\nu = g_{\mu\nu}dx^\mu dx^\nu$$

or

$$\left(g'_{\alpha\beta}\frac{\partial x'^\alpha}{\partial x^\mu}\frac{\partial x'^\beta}{\partial x^\nu} - g_{\mu\nu}\right)dx^\mu dx^\nu = 0.$$

The above equation is identically zero for arbitrary dx^μ, so we have

$$g_{\mu\nu} = \frac{\partial x'^\alpha}{\partial x^\mu}\frac{\partial x'^\beta}{\partial x^\nu}g'_{\alpha\beta} \tag{2.10}$$

that is, $g_{\mu\nu}$ is a covariant tensor of rank two. It is called the *metric tensor* or the *fundamental tensor*. The metric tensor is locally Minkoskian.

So far, covariant and contravariant vectors have no direct connection with each other except that their inner product is an invariant. A space in which covariant and contravariant vectors exist separately is called *affine*. Physical quantities are independent of the particular choice of the mode of description, that is, independent of the possible choices of contravariance or covariance. In metric space, contravariant and covariant vectors can be converted into each other with the help of the fundamental tensor $g_{\mu\nu}$. For example, we can get the covariant vector A_μ from the contravariant vector A^ν

$$A_\mu = g_{\mu\nu}A^\nu. \tag{2.11}$$

Since the determinant $|g|$ does not vanish, these equations can be solved for A^ν in terms of the A_μ. Let the result be

$$A^\nu = g^{\mu\nu}A_\mu \tag{2.12}$$

By combining the two transformations (2.11) and (2.12), we have

$$A_\mu = g_{\mu\nu}g^{\nu\alpha}A_\alpha.$$

Since the equation must hold for any four quantities A_μ, we can infer

$$g_{\mu\nu}g^{\nu\alpha} = \delta_\mu{}^\alpha. \tag{2.13}$$

In other words, $g^{\mu\nu}$ is the inverse of $g_{\mu\nu}$ and vice versa, that is

$$g^{\mu\nu} = \frac{M^{\mu\nu}}{|g|} \tag{2.14}$$

where $M^{\mu\nu}$ is the minor of the element $g_{\mu\nu}$.

Equation (2.11) may be used to lower any upper index occurring in a tensor. Similarly, (2.12) can be used to raise any downstairs index. It is necessary to remember the position from which the index was lowered or raised, because when we bring the index back to its original site, we do not want to interchange the order of indexes, in general $T^{\mu\nu} \neq T^{\nu\mu}$.

Two tensors, $A_{\mu\nu}$ and $B^{\mu\nu}$, are said to be reciprocal to each other if

$$A_{\mu\nu} B^{\nu\alpha} = \delta^{\alpha}{}_{\mu}. \tag{2.15}$$

A tensor is called symmetric with respect to two contravariant or two covariant indexes if its components remain unaltered on interchange of the indexes. For example, if $A^{\mu\nu\alpha}_{\beta\gamma} = A^{\nu\mu\alpha}_{\beta\gamma}$, the tensor is symmetric in μ and ν. If a tensor is symmetric with respect to any two contravariant and any two covariant indexes, it is called symmetric.

A tensor is called skew-symmetric with respect to two contravariant or two covariant indexes if its components change sign upon interchange of the indexes. Thus, if $A^{\mu\nu\alpha}_{\beta\gamma} = -A^{\nu\mu\alpha}_{\beta\gamma}$, the tensor is skew-symmetric in μ and ν.

If a tensor is symmetric (or skew-symmetric) with respect to two indexes in one coordinate system, it remains symmetric (skew-symmetric) with respect to those two indexes in any other coordinate system. It is easy to prove this. For example, if $B^{\alpha\beta}$ is symmetric, $B^{\alpha\beta} = B^{\beta\alpha}$, then

$$B'^{\alpha\beta} = \frac{\partial x'^{\alpha}}{\partial x^{\gamma}} \frac{\partial x'^{\beta}}{\partial x^{\delta}} B^{\gamma\delta} = \frac{\partial x'^{\alpha}}{\partial x^{\gamma}} \frac{\partial x'^{\beta}}{\partial x^{\delta}} B^{\delta\gamma} = B'^{\beta\alpha} \tag{2.16}$$

i.e., the tensor remains symmetric in the primed coordinate system.

Every tensor can be expressed as the sum of two tensors, one of which is symmetric and the other skew-symmetric in a pair of covariant or contravariant indices. Consider, for example, the tensor $B^{\alpha\beta}$. We can write it as

$$B^{\alpha\beta} = 1/2(B^{\alpha\beta} + B^{\beta\alpha}) + 1/2(B^{\alpha\beta} - B^{\beta\alpha}) \tag{2.17}$$

with the first term on the right-hand side symmetric and the second term skew-symmetric. By similar reasoning the result is seen to be true for any tensor.

It is obvious that the sum or difference of two or more tensors of the same rank and type (i.e., same number of contravariant indices and same number of covariant indices) is also a tensor of the same rank and type. Thus if $A_{\lambda}{}^{\mu\nu}$ and $B_{\lambda}{}^{\mu\nu}$ are tensors, then $C_{\lambda}{}^{\mu\nu} = A_{\lambda}{}^{\mu\nu} + B_{\lambda}{}^{\mu\nu}$ and $D_{\lambda}{}^{\mu\nu} = A_{\lambda}{}^{\mu\nu} - B_{\lambda}{}^{\mu\nu}$ are also tensors.

A given quantity $N^{\mu\ldots}_{\nu\rho\ldots}$ with various up and down indexes may or may not be a tensor. We can test whether it is a tensor or not by using the quotient law, which can be stated as follows:

Suppose we have a quantity X and we do not know whether it is a tensor or not. If an inner product of X with an arbitrary tensor is a tensor, then X is also a tensor.

For example, let $X = P_{\lambda\mu\nu}$, and A^{λ} is an arbitrary contravariant vector; if $P_{\lambda\mu\nu} A^{\lambda} = Q_{\mu\nu}$ is a tensor, then $P_{\lambda\mu\nu}$ is a contravariant tensor of rank 3. We can prove this explicitly:

$$A^\lambda P_{\lambda\mu\nu} = \frac{\partial x'^\alpha}{\partial x^\mu}\frac{\partial x'^\beta}{\partial x^\nu}A'^\gamma P'_{\gamma\alpha\beta}.$$

but

$$A'^\gamma = \frac{\partial x'^\gamma}{\partial x^\lambda}A^\lambda.$$

Hence,

$$A^\lambda P_{\lambda\mu\nu} = \frac{\partial x'^\alpha}{\partial x^\mu}\frac{\partial x'^\beta}{\partial x^\nu}\frac{\partial x'^\gamma}{\partial x^\lambda}A^\lambda P'_{\gamma\alpha\beta}.$$

This equation must hold for all values of A^λ, so we have

$$P_{\lambda\mu\nu} = \frac{\partial x'^\alpha}{\partial x^\mu}\frac{\partial x'^\beta}{\partial x^\nu}\frac{\partial x'^\gamma}{\partial x^\lambda}P'_{\gamma\alpha\beta}, \tag{2.18}$$

showing that $P_{\lambda\mu\nu}$ is a contravariant tensor of rank 3.

For a nontensor $N^{\mu\cdots}_{\nu\rho\ldots}$ we can raise and lower indexes by the same rules as for a tensor. Thus, for example

$$g^{a\nu}N^\rho_{\nu\eta} = N^{a\mu}_\eta. \tag{2.19}$$

2.2 Parallel Displacement and Covariant Differentiation

In this section and the following three sections we will give the full of apparatus of differential geometry. The reader may be in danger of being overwhelmed by algebra, but to simplify mathematics is not to make it simple, either. Please do not despair; just relax and try to enjoy it.

We have seen that a covariant vector is transformed according to the formula

$$A_i = \frac{\partial x'^k}{\partial x^i}A'_k \tag{2.20}$$

where the coefficients are functions of the coordinates. So vectors at different points transform differently. Because of this fact, dA_i is not a vector, since it is the difference of two vectors located at two infinitesimally separated points of space-time. We can easily verify this directly from (2.20)

$$dA_i = \frac{\partial x'^k}{\partial x^i}dA'_k + A'_k\frac{\partial^2 x'^k}{\partial x^i\partial x^j}dx^j$$

which shows that dA_i does not transform at all like a vector. The same also applies to the differential of a contravariant vector.

When using curvilinear coordinates, a differential can be obtained only when the two vectors to be subtracted from each other are located at the same point in space-time. In order to do so, we must what we call *parallel displace* one of the vectors to the point where the other vector is located, after which we determine the difference of two vectors, which now refer to one and the same point in space-time.

The concept of parallel displacement of a vector is very clear in Cartesian co-ordinates: displace a vector parallel to itself so that both its length and orientation are unchanged. We can extend the idea of parallel displacement of a vector to curved spaces in a consistent way. This requires us to assume that there always exist Galilean coordinates in the immediate vicinity of a point in space-time; in such a coordinate system the idea of an infinitesimal parallel displacement of a vector works. In other words, a vector can be transported parallel to itself without changing its length and orientation. We illustrate, in Fig. 2.1, with the example of a curved two-dimensional surface in a three-dimensional Euclidean space. During the infinitesimal parallel displacement of two vectors, A^μ and B^μ, the angle between them clearly remains unchanged, and so the inner (scalar) product of two vectors, $A_\mu B^\mu$, does not change under parallel displacement. For arbitrary coordinates we define the operation of infinitesimal parallel displacement of a vector A^μ from Point P to a neighboring Point Q to be one that leaves the inner product with an arbitrary vector B^μ invariant.

Parallel displacements are independent of the paths taken on a Euclidean plane (a flat surface), as shown in Fig. 2.2a. On a curved surface, however, we will obtain a different final result on the path taken (Figure 2.2b).

We can transfer a vector continuously along a path by the process of parallel displacement. In curvilinear coordinates, the components of a vector would be expected

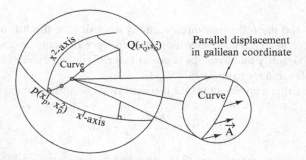

Fig. 2.1 Parallel displacement in curvilinear coordinates.

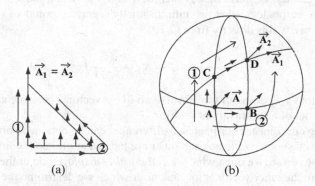

Fig. 2.2 Parallel transport around a closed curve.

to change under a parallel displacement, unlike the case of a Cartesian coordinate. Therefore, if A^μ are the components of a contravariant vector at the point $P(x^\mu)$, and $A^\mu + dA^\mu$ the components at a neighboring point $Q(x^\mu + dx^\mu)$, where

$$A^\mu + dA^\mu = A^\mu + \frac{\partial A^\mu}{\partial x^\sigma} dx^\sigma \tag{2.21}$$

an infinitesimal parallel displacement of A^μ from P to Q would produce a variation of its components, δA^μ. δA^μ should be a linear function of the coordinate differentials and the components A^μ. We write it in the form

$$\delta A^\nu = -\Gamma^\nu_{\alpha\beta} A^\alpha dx^\beta \tag{2.22}$$

where the $\Gamma^\nu_{\alpha\beta}$ are certain functions of the coordinates and are called Christoffel symbols of the second kind. Their form depends on the coordinate system. It will be proved in Section 4.4 that in a Galilean coordinate system $\Gamma^\nu_{\alpha\beta} = 0$. From this it is already clear that the quantities $\Gamma^\nu_{\alpha\beta}$ do not form a tensor, since a tensor that is equal to zero in one coordinate system is equal to zero in every other one. In a curved space it is impossible to make all the $\Gamma^\nu_{\alpha\beta}$ vanish over all of space.

The vector resulting from parallel displacement from Point P to Point Q is $A^\mu + \delta A^\mu$. Subtraction of these two quantities gives us

$$DA^\mu = dA^\mu - \delta A^\mu = \left(\frac{\partial A^\mu}{\partial x^\sigma} + \Gamma^\mu_{\sigma\alpha} A^\alpha \right) dx^\sigma. \tag{2.23}$$

We would expect the difference $dA^\mu - \delta A^\mu$ to be a vector since it is the difference of two vectors at the same point; the quantity

$$\frac{\partial A^\mu}{\partial x^\sigma} + \Gamma^\mu_{\sigma\alpha} A^\alpha$$

then is a mixed tensor called the covariant derivative of A^μ and written

$$A^\mu_{;\sigma} = \frac{\partial A^\mu}{\partial x^\sigma} + \Gamma^\mu_{\sigma\alpha} A^\alpha. \tag{2.24}$$

From $\delta(A_\mu A^\mu) = 0$ it follows, using (2.22), that

$$\delta A_\mu = \Gamma^\alpha_{\mu\beta} A_\alpha dx^\beta.$$

From this and a similar procedure that leads to (2.23) and (2.24), we obtain the covariant derivative of A_μ:

$$A_{\mu;\sigma} = \frac{\partial A_\mu}{\partial x^\sigma} - \Gamma^\alpha_{\mu\sigma} A_\alpha. \tag{2.25}$$

The tensor character of (2.24) and (2.25) can be established formally by showing that they obey the required transformation laws. This will require us first to establish the transformation laws for the $\Gamma^\alpha_{\mu\sigma}$. It is not difficult to do this, but very tedious, and so we shall not do it in this book.

To obtain the contravariant derivative, we raise the index that denotes differentiation,

$$A^{\mu;\sigma} = g^{\sigma\,\alpha} A^{\mu}_{;\alpha}. \tag{2.26}$$

In Galilean coordinates, $\Gamma^{\alpha}_{\mu\sigma} = 0$, and so covariant differentiation reduces to ordinary differentiation.

We may also obtain the covariant derivative of a tensor by determining the change in the tensor under an infinitesimal parallel displacement. For example, let us consider any arbitrary tensor $T^{\mu\nu}$ expressible as a product of two contravariant vectors $A^{\mu} B^{\nu}$. Under infinitesimal parallel displacement

$$\delta(A^{\mu} B^{\nu}) = A^{\mu}\delta B^{\nu} + B^{\nu}\delta A^{\mu} = -A^{\mu}\Gamma^{\nu}_{\alpha\beta} B^{\alpha} dx^{\beta} - B^{\nu}\Gamma^{\mu}_{\beta\sigma} A^{\sigma} dx^{\beta}.$$

By virtue of the linearity of this transformation we also have

$$\delta A^{\mu\nu} = -\left(A^{\mu\beta}\Gamma^{\nu}_{\beta\alpha} + A^{\beta\nu}\Gamma^{\mu}_{\beta\alpha}\right) dx^{\alpha}.$$

Substituting this in

$$DA^{\mu\nu} = dA^{\mu\nu} - \delta A^{\mu\nu} = A^{\mu\nu}_{;\alpha} dx^{\alpha}$$

we get the covariant derivative of the tensor $T^{\mu\nu}$ in the form

$$T^{\mu\nu}_{;\alpha} = \frac{\partial T^{\mu\nu}}{\partial x^{\alpha}} + \Gamma^{\mu}_{\beta\alpha} T^{\beta\nu} + \Gamma^{\nu}_{\beta\alpha} T^{\mu\beta}. \tag{2.27}$$

In similar fashion we obtain the covariant derivative of the mixed tensor T^{μ}_{ν} and the covariant tensor $T_{\nu\mu}$ in the form

$$T^{\mu}_{\nu;\alpha} = \frac{\partial T^{\mu}_{\nu}}{\partial x^{\alpha}} - \Gamma^{\beta}_{\nu\alpha} T^{\mu}_{\beta} + \Gamma^{\mu}_{\beta\alpha} T^{\beta}_{\nu}, \tag{2.28}$$

$$T_{\mu\nu;\alpha} = \frac{\partial T_{\mu\nu}}{\partial x^{\alpha}} - \Gamma^{\beta}_{\mu\alpha} T_{\beta\nu} - \Gamma^{\beta}_{\nu\alpha} T_{\mu\beta}. \tag{2.29}$$

One can similarly determine the covariant derivative of a tensor of arbitrary rank. In doing this one finds the following rule of covariant differentiation:

To obtain the covariant derivative of the tensor $T^{\cdots}_{\cdots}$ with respect to x^{μ}, you add to the ordinary derivative $\partial T^{\cdots}_{\cdots}/\partial x^{\mu}$ for each covariant index $v(T^{\cdots}_{\cdots v\cdots})$ a term $-\Gamma^{\alpha}_{\mu v} T^{\cdots}_{\cdots\alpha\cdots}$, and for each contravariant index $v(T^{\cdots v\cdots}_{\cdots})$ a term $+\Gamma^{v}_{\alpha\mu} T^{\cdots\alpha\cdots}_{\cdots}$.

The covariant derivative of the metric tensor $g_{\mu\nu}$ is zero. To show this we note that the relation

$$DA_{\mu} = g_{\mu\nu} DA^{\nu}$$

is valid for the vector DA_{μ} as for any vector. On the other hand, we have $A_{\mu} = g_{\mu\nu} A^{\nu}$, so that

$$DA_{\mu} = D(g_{\mu\nu} A^{\nu}) = g_{\mu\nu} DA^{\nu} + A^{\nu} Dg_{\mu\nu}.$$

Comparing with $DA_{\mu} = g_{\mu\nu} DA^{\nu}$, we have $A^{\nu} Dg_{\mu\nu} = 0$. But the vector A^{ν} is arbitrary, so

$$Dg_{\mu\nu} = 0.$$

Therefore, the covariant derivative is

$$g_{\mu\nu;\alpha} = 0. \tag{2.30}$$

Thus, $g_{\mu\nu}$ may be considered as a constant during covariant differentiation.

The covariant derivative of a product can be found by the same rule as for ordinary differentiation of products. In doing this we must consider the covariant derivative of a scalar ϕ as an ordinary derivative, that is, as the covariant vector $\phi_k = \partial\phi/\partial x^k$, in accordance with the fact that for a scalar $\delta\phi = 0$, and hence $D\phi = d\phi$. For example

$$(A_\mu B_\nu)_{;\alpha} = A_{\mu;\alpha}B_\nu + A_\mu B_{\nu;\alpha}. \tag{2.31}$$

2.3 Symmetry Properties of the Christoffel Symbols

We now show that $\Gamma^\nu_{\alpha\beta}$ is symmetric in the subscripts. If δA^ν is a coordinate differential dx^ν, then (2.22) becomes

$$\delta(dx^\nu) = -\Gamma^\nu_{\alpha\beta}dx^\alpha dx^\beta. \tag{2.32}$$

Next, we return to the local Cartesian coordinate system by the transformations

$$\left. \begin{array}{l} x^\alpha = f^\alpha(x'^1, x'^2, \dots.) \\ x'^\alpha = \varphi^\alpha(x^1, x^2, \dots\dots) \end{array} \right\} \tag{2.33}$$

where the primed coordinates are local Cartesian coordinates. From (2.33) we obtain

$$dx^\alpha = \frac{\partial f^\alpha}{\partial x'^\beta}dx'^\beta. \tag{2.34}$$

Under a parallel displacement, $\delta(dx'^\beta) = 0$, so that from (2.34) we have

$$\delta(dx^\nu) = \frac{\partial^2 f^\nu}{\partial x'^\delta \partial x'^\gamma}dx'^\delta dx'^\gamma = \frac{\partial^2 f^\nu}{\partial x'^\delta \partial x'^\gamma}\frac{\partial \varphi^\delta}{\partial x^\alpha}\frac{\partial \varphi^\gamma}{\partial x^\beta}dx^\alpha dx^\beta.$$

Comparing this with (2.32) we obtain

$$\Gamma^\nu_{\alpha\beta} = -\frac{\partial^2 f^\nu}{\partial x'^\delta \partial x'^\gamma}\frac{\partial \varphi^\delta}{\partial x^\alpha}\frac{\partial \varphi^\gamma}{\partial x^\beta}. \tag{2.35}$$

The right-hand side is clearly symmetric in the indexes α and β, so that $\Gamma^\nu_{\alpha\beta}$ is also symmetric in α and β.

2.4 Christoffel Symbols and the Metric Tensor

It is very useful to express the Γ's in terms of the metric tensors. Let A^μ be any contravariant vector, $A_\mu = g_{\mu\nu} A^\nu$ a covariant vector. From the definition of parallel displacement $\delta(A_\mu A^\mu) = 0$, we have

$$\delta(A_\mu A^\mu) = g_{\mu\nu}(x^\mu + dx^\mu)[A^\nu + \delta A^\nu][A^\mu + \delta A^\mu] - g_{\mu\nu}(x^\mu)A^\nu A^\mu = 0.$$

Carrying out these operations gives us

$$\frac{\partial g_{\mu\nu}}{\partial x^\alpha} A^\mu A^\nu dx^\alpha + g_{\mu\nu} A^\mu \delta A^\nu + g_{\mu\nu} A^\nu \delta A^\mu = 0.$$

Making use of (2.22) to eliminate δA^μ and δA^ν gives us

$$\frac{\partial g_{\mu\nu}}{\partial x^\alpha} - g_{\nu\beta} \Gamma^\beta_{\nu\alpha} - g_{\nu\beta} \Gamma^\beta_{\mu\alpha} = 0. \tag{2.36}$$

Now, $\Gamma^\beta_{\nu\alpha}$ is symmetric in the lower indexes ν and α, and this symmetry allows permutation of ν and α to obtain

$$\frac{\partial g_{\mu\alpha}}{\partial x^\nu} - g_{\mu\beta} \Gamma^\beta_{\nu\alpha} - g_{\alpha\beta} \Gamma^\beta_{\mu\nu} = 0. \tag{2.37}$$

Similarly, we write

$$\frac{\partial g_{\nu\alpha}}{\partial x^\mu} - g_{\nu\beta} \Gamma^\beta_{\mu\alpha} - g_{\alpha\beta} \Gamma^\beta_{\mu\nu} = 0. \tag{2.38}$$

Solving (2.36), (2.37), (2.38) for $\Gamma^\gamma_{\mu\alpha}$, we obtain

$$\Gamma^\gamma_{\mu\alpha} = \frac{1}{2} g^{\gamma\nu} \left[\frac{\partial g_{\nu\mu}}{\partial x^\alpha} + \frac{\partial g_{\nu\alpha}}{\partial x^\mu} - \frac{\partial g_{\mu\alpha}}{\partial x^\nu} \right]. \tag{2.39}$$

The Christoffel symbol of the first kind is

$$\Gamma_{\nu,\mu\alpha} = \frac{1}{2} \left[\frac{\partial g_{\nu\mu}}{\partial x^\alpha} + \frac{\partial g_{\nu\alpha}}{\partial x^\mu} - \frac{\partial g_{\nu\alpha}}{\partial x^\nu} \right]. \tag{2.40}$$

It is often written as $[\mu\alpha, \nu]$. Clearly $\Gamma_{\nu,\mu\alpha} = \Gamma_{\nu,\alpha\mu}$. The Christoffel symbols are also known as the *affine connections*. The Christoffel symbols all vanish in Galilean coordinates, as the metric tensors are all constants in Galilean coordinates.

The equation $g_{\mu\nu;\sigma} = 0$ can be used to offer an alternative derivation of (2.39) and (2.40). We write in accordance with the general definition (2.29):

$$g_{\mu\nu;\alpha} = \frac{\partial g_{\mu\nu}}{\partial x^\alpha} - g_{\beta\nu} \Gamma^\beta_{\mu\alpha} - g_{\mu\beta} \Gamma^\beta_{\nu\alpha} = \frac{\partial g_{\mu\nu}}{\partial x^\alpha} - \Gamma_{\nu,\mu\alpha} - \Gamma_{\mu,\nu\alpha} = 0.$$

From this we have, permuting the indexes μ, ν, α:

$$\frac{\partial g_{\mu\nu}}{\partial x^\alpha} = \Gamma_{\nu,\mu\alpha} + \Gamma_{\mu,\nu\alpha}.$$

$$\frac{\partial g_{\alpha\mu}}{\partial x^\nu} = \Gamma_{\mu,\nu\alpha} + \Gamma_{\alpha,\mu\nu}.$$

$$\frac{\partial g_{\nu\alpha}}{\partial x^\mu} = \Gamma_{-\alpha,\nu\mu} - \Gamma_{\nu,\alpha\mu}.$$

Taking half the sum of these equations and remembering that $\Gamma_{\mu,\nu\alpha} = \Gamma_{\mu,\alpha\nu}$, we find

$$\Gamma_{\mu,\nu\alpha} = \frac{1}{2}\left(\frac{\partial g_{\mu\nu}}{\partial x^\alpha} + \frac{\partial g_{\mu\alpha}}{\partial x^\nu} - \frac{\partial g_{\nu\alpha}}{\partial x^\mu} \right). \tag{2.41}$$

From this we have for the symbols $\Gamma^\mu_{\nu\alpha} = g^{\nu\beta}\Gamma_{\mu,\nu\alpha}$

$$\Gamma^\mu_{\nu\alpha} = \frac{1}{2}g^{\mu\beta}\left(\frac{\partial g_{\beta\nu}}{\partial x^\alpha} + \frac{\partial g_{\beta\alpha}}{\partial x^\nu} - \frac{\partial g_{\nu\alpha}}{\partial x^\beta} \right). \tag{2.42}$$

A coordinate system in which the Christoffel symbols vanish at Point P is called a *geodesic coordinate system*, and Point P is said to be the *pole*.

2.5 Geodesics

As an application of the notion of parallel displacement and covariant differentiation, let's consider the geodesic equation. A geodesic is the curve defined by the requirement that each element of it is a parallel displacement of the preceding element. We shall see later that the world line of a point-like particle not acted upon by any forces, except gravitation, is a time-like geodesic.

If we take a point with coordinates x^μ and move it along a path, we then have x^μ as a function of some parameter s. There is a tangent vector $t^\mu = dx^\mu/ds$ at each point of the path. As we go along the path the vector t^μ gets shifted by parallel displacement: we shift the initial position from x^μ to $x^\mu + t^\mu ds$, and then shift the vector t^μ to this new position by parallel displacement, then shift the point again in the direction fixed by the new t^μ, and so on. If we are given the initial point and the initial value of the vector t^μ, not only can the path be determined but also the parameter s along it. A path produced in this way is called a geodesic. We get the geodesic equations by applying (2.22) with $A^\mu = t^\mu$

$$\frac{dt^\nu}{ds} + \Gamma^\nu_{\mu\sigma}t^\mu\frac{dx^\sigma}{ds} = 0, \tag{2.43}$$

or

$$\frac{d^2x^\nu}{ds^2} + \Gamma^\nu_{\mu\sigma}\frac{dx^\sigma}{ds}\frac{dx^\mu}{ds} = 0. \tag{2.44}$$

If the vector t^μ is initially a null vector, it always remains a null vector and the path is called a *null geodesic*. If the vector t^μ is initially time-like (i.e., $t^\mu t^\mu > 0$), it is always time-like and we have a time-like geodesic. If t^μ is initially space-like ($t^\mu t^\mu < 0$), it is always space-like and we have a space-like geodesic.

For a time-like geodesic we may multiply the initial t^μ by a factor so as to make its length unity. This requires only a change in the scale of s. The vector t^μ now always has a unit length. It is simply the velocity vector $u^\mu = dx^\mu/d\tau$, and the parameter s has becomes the proper time τ. (2.43) becomes

$$\frac{du^\mu}{d\tau} + \Gamma^\nu_{\mu\sigma} u^\mu u^\sigma = 0 \tag{2.43a}$$

and (2.44) becomes

$$\frac{d^2 x^\nu}{d\tau^2} + \Gamma^\nu_{\mu\sigma} \frac{dx^\sigma}{d\tau}\frac{dx^\mu}{d\tau} = 0. \tag{2.44a}$$

We make the physical assumption that the world line of a particle not acted on by any forces, except gravitation, is a time-like geodesic. Note that the terms $-m\Gamma^\nu_{\mu\sigma} u^\mu u^\sigma$ in (2.43a) may be interpreted as the gravitational forces, and the components of the metric tensor $g_{\mu\nu}$ play the role of the classical gravitational potential (as the Christoffel symbol is proportional to the derivatives of the metric tensor; see [2.39] and [2.40]).

Choosing a local Galilean frame in which $g_{\mu\nu} = $ constants, the Christoffel symbols $\Gamma^\nu_{\mu\sigma} = 0$, and $du^\mu/d\tau = 0$. Therefore, the gravitational forces can be locally eliminated, and the geodesic equations can be locally reduced to the special relativistic equations of motion, in agreement with the equivalence.

The path of a light ray is a null geodesic. It is fixed by (2.44) referring to some parameter s along the path. The proper time τ cannot now be used because $d\tau$ vanishes.

2.6 The Stationary Property of Geodesics

We now examine the stationary property of geodesics. A geodesic that is not a null geodesic joining two points P and Q has a stationary value compared with the interval (line element) measured along another neighboring curve joining P and Q. This property holds good for a straight line in flat space and, in that case, it is also true that the straight line gives the shortest interval from one point to another. In curved space, the geodesic is no longer a straight line because space-time is no longer flat, and the particle motion is not rectilinear and uniform, in general. However, we can show that the geodesic is a path of extreme length. (We will not enquire whether or not the geodesic in a curved space gives the minimum or maximum value of the interval between any of its points). To show this, we demonstrate that the relations which must be satisfied to give a stationary value to the integral

$$s = \int \sqrt{g_{\lambda\mu} dx^\lambda dx^\mu} \tag{2.45}$$

are simply the equations of geodesics (2.44) of the previous section. Let us first introduce a parameter α and write (2.45) as

$$s = \int_{\alpha_P}^{\alpha_Q} \left[g_{\lambda\mu} \frac{dx^\lambda}{d\alpha} \frac{dx^\mu}{d\alpha} \right]^{1/2} d\alpha$$

where α varies from point to point of the geodesic curve described by the relations which we are seeking for, and we write it as

$$x^\mu = f^\mu(\alpha). \tag{2.46}$$

Any other neighboring curve joining P and Q has equations of the form

$$\bar{x}^\lambda = x^\lambda + \varepsilon y^\lambda = f^\lambda(\alpha) + \varepsilon y^\lambda(\alpha)$$

where $y^\lambda = 0$ at the end points P and Q, i.e., at $\alpha = \alpha_P$ and $\alpha = \alpha_Q$, and ε is a small quantity whose square and higher powers are negligibly small. If $\bar{s}$ is the line element along the neighboring curve joining P and Q, then

$$\bar{s} = \int_{\alpha_P}^{\alpha_Q} \left(g_{\lambda\mu}(\bar{x}) \frac{d\bar{x}^\lambda}{d\alpha} \frac{d\bar{x}^\mu}{d\alpha} \right)^{1/2} d\alpha$$

and therefore, neglecting all powers of ε higher than the first,

$$\bar{s} - s = \int_{\alpha_P}^{\alpha_Q} \left[g_{\lambda\mu} \frac{dx^\lambda}{d\alpha} \frac{dx^\mu}{d\alpha} + \varepsilon \left(\frac{\partial g_{\lambda\mu}}{dx^\sigma} \frac{dx^\lambda}{d\alpha} y^\sigma + 2 g_{\lambda\mu} \frac{dy^\lambda}{d\alpha} \right) \frac{dx^\mu}{d\alpha} \right]^{1/2} d\alpha$$
$$- \int_{\alpha_P}^{\alpha_O} \left[g_{\lambda\mu} \frac{dx^\lambda}{d\alpha} \frac{dx^\mu}{d\alpha} \right]^{1/2} d\alpha,$$

which can be reduced to

$$\bar{s} - s = \frac{1}{2}\varepsilon \int_{\alpha_P}^{\alpha_Q} \left[\frac{\partial g_{\lambda\mu}}{dx^\sigma} \frac{dx^\lambda}{d\alpha} y^\sigma + 2 g_{\lambda\mu} \frac{dy^\lambda}{d\alpha} \right] \frac{dx^\mu}{d\alpha} \frac{d\alpha}{ds} ds.$$

Note that $ds = \sqrt{g_{\lambda\mu} dx^\lambda dx^\mu}$.

We can simplify the calculation, if $s \neq 0$, by assuming that the parameter α is identical with s itself measured along the geodesic. Then $d\alpha/ds = 1$ and

$$\bar{s} - s = \frac{1}{2}\varepsilon \int_{\alpha_P}^{\alpha_Q} \left[\frac{\partial g_{\lambda\mu}}{dx^\sigma} \frac{dx^\lambda}{d\alpha} y^\sigma + 2 g_{\lambda\mu} \frac{dy^\lambda}{d\alpha} \right] \frac{dx^\mu}{ds} ds,$$

with the x^λ, y^λ now regarded as functions of s. Integration of the second term in the last equation by parts yields

$$\bar{s} - s = \frac{1}{2}\varepsilon \int_{\alpha_P}^{\alpha_Q} \left[\frac{\partial g_{\lambda\mu}}{dx^\sigma} \frac{dx^\lambda}{ds} \frac{dx^\mu}{ds} - 2 \frac{d}{ds} \left(g_{\sigma\mu} \frac{dx^\mu}{ds} \right) \right] y^\sigma ds + \varepsilon \left(g_{\lambda\mu} \frac{dx^\mu}{ds} y^\lambda \right)_{s_P}^{s_Q}.$$

But the functions y^λ vanish at s_P and s_Q; hence the integrated term is zero. Therefore, if the interval is to have a stationary value for the geodesic curve compared with neighboring curves, $\bar{s} - s$ must be zero for any choice of the function y^λ. This is possible only if the coefficient of each y^σ in the integrand is separately zero, and therefore the differential equations of the geodesic are the n equations

$$\frac{d}{ds}\left(g_{\sigma\mu}\frac{dx^\mu}{ds}\right) - \frac{1}{2}\frac{\partial g_{\lambda\mu}}{\partial x^\sigma}\frac{dx^\lambda}{ds}\frac{dx^\mu}{ds} = 0 \quad (\sigma = 1, 2, \ldots, n). \tag{2.47}$$

Now,

$$\frac{d}{ds}\left(g_{\sigma\mu}\frac{dx^\mu}{ds}\right) - g_{\sigma\mu}\frac{d^2x^\mu}{ds^2} + \frac{\partial g_{\mu\sigma}}{\partial x^\lambda}\frac{dx^\lambda}{ds}\frac{dx^\mu}{ds}$$

$$= g_{\sigma\mu}\frac{d^2x^\mu}{ds^2} + \frac{1}{2}(g_{\mu\sigma,\lambda} + g_{\lambda\sigma,\mu})\frac{dx^\lambda}{ds}\frac{dx^\mu}{ds}.$$

Thus, the equations of the geodesic may be written

$$g_{\sigma\mu}\frac{d^2x^\mu}{ds^2} + \frac{1}{2}(g_{\mu\sigma,\lambda} + g_{\lambda\sigma,\mu} - g_{\lambda\mu,\sigma})\frac{dx^\lambda}{ds}\frac{dx^\mu}{ds}$$

$$= g_{\sigma\mu}\frac{d^2x^\mu}{ds^2} + \Gamma_{\sigma,\lambda\mu}\frac{dx^\lambda}{ds}\frac{dx^\mu}{ds} = 0. \tag{2.48}$$

Multiplying this by $g^{\tau\sigma}$ and summing over σ, the result is

$$\frac{d^2x^\tau}{ds^2} + \Gamma^\mu_{\lambda\mu}\frac{dx^\lambda}{ds}\frac{dx^\mu}{ds} = 0, \quad (\tau = 1, 2, \ldots, n), \tag{2.49}$$

which is simply the standard form of (2.44) for geodesics.

The above work shows that we may use the stationary condition as the definition of a geodesic, except in dealing with the propagation of light. In that case, we have null geodesics, so the deduction as given above cannot be applied because ds vanish throughout.

2.7 The Curvature Tensor

In a flat space, if we perform two (ordinary) differentiations in succession their order does not matter. However, this does not, in general, hold for covariant differentiation in a curved space, except for a scalar ϕ. For the case of a scalar, we have

$$\phi_{;\mu;\nu} = (\phi_{;\mu})_{,\nu} - \Gamma^\alpha_{\mu\nu}\phi_{;\alpha} = \phi_{,\mu,\nu} - \Gamma^\alpha_{\mu\nu}\phi_{,\sigma}. \tag{2.50}$$

Since $\Gamma^\alpha_{\mu\nu}$ is symmetric in the lower indexes μ and ν, so the order of differentiation does not matter.

Now if we take a contravariant vector A^μ and apply covariant differentiations twice to it, we will find the order of differentiation is very important. First, the covariant differentiation of A^μ gives a mixed tensor

$$A^\mu_{;\nu} = \frac{\partial A^\mu}{\partial x^\nu} + \Gamma^\mu_{\alpha\nu} A^\alpha$$

Covariant differentiation of this mixed tensor gives

$$A^\mu_{;\nu;\beta} = \frac{\partial}{\partial x^\beta}\left(A^\mu_{;\nu}\right) + \Gamma^\mu_{\alpha\beta} A^\alpha_{;\nu} - \Gamma^\alpha_{\nu\beta} A^\mu_{;\alpha}$$

or

$$A^\mu_{;\nu;\beta} = \frac{\partial^2 A^\mu}{\partial x^\beta \partial x^\nu} + \Gamma^\mu_{\alpha\nu}\frac{\partial A^\alpha}{\partial x^\beta} + A^\alpha\frac{\partial \Gamma^\mu_{\alpha\nu}}{\partial x^\beta} + A^\alpha_{;\nu}\Gamma^\mu_{\alpha\beta} - A^\mu_{;\alpha}\Gamma^\alpha_{\beta\nu}. \tag{2.51}$$

Interchanging β and ν, we obtain

$$A^\mu_{;\beta;\nu} = \frac{\partial^2 A^\mu}{\partial x^\nu \partial x^\beta} + \Gamma^\mu_{\alpha\beta}\frac{\partial A^\alpha}{\partial x^\nu} + A^\alpha\frac{\partial \Gamma^\mu_{\alpha\beta}}{\partial x^\nu} + A^\alpha_{;\beta}\Gamma^\mu_{\alpha\nu} - A^\mu_{;\alpha}\Gamma^\alpha_{\beta\nu} \tag{2.52}$$

Subtracting (2.52) from (2.51), we get

$$A^\mu_{;\nu;\beta} - A^\mu_{;\beta;\nu} = A^\alpha\left[\frac{\partial \Gamma^\mu_{\alpha\nu}}{\partial x^\beta} - \frac{\partial \Gamma^\mu_{\alpha\beta}}{\partial x^\nu} + \Gamma^\gamma_{\alpha\nu}\Gamma^\mu_{\gamma\beta} - \Gamma^\gamma_{\alpha\beta}\Gamma^\mu_{\gamma\nu}\right] = A^\alpha R^\mu_{\alpha\nu\beta} \tag{2.53}$$

where

$$R^\mu_{\alpha\nu\beta} = \frac{\partial \Gamma^\mu_{\alpha\nu}}{\partial x^\beta} - \frac{\partial \Gamma^\mu_{\alpha\beta}}{\partial x^\nu} + \Gamma^\gamma_{\alpha\nu}\Gamma^\mu_{\gamma\beta} - \Gamma^\gamma_{\alpha\beta}\Gamma^\mu_{\gamma\nu}. \tag{2.54}$$

Since $A^\mu_{;\nu;\beta} - A^\mu_{;\beta;\nu}$ and A^α are tensors, $R^\mu_{\alpha\nu\beta}$ must be the component of a tensor, by the quotient law. It is called the *curvature tensor* or the *Riemann tensor*, and it depends solely on the Christoffel symbols and their derivatives. In flat space all $g_{\mu\nu}$ can be transformed into constants (rectangular or Galilean coordinates) and all Christoffel symbols vanish; hence $R^\mu_{\alpha\nu\beta} = 0$. Being a tensor equation, it holds in all coordinate systems (Cartesian, oblique, or curvilinear). In a curved space, $R^\mu_{\alpha\nu\beta}$ will not vanish. Therefore it is a measure of the curvature of space.

From (2.54) it follows that the curvature tensor is antisymmetric in the indices ν and β:

$$R^\mu_{\alpha\nu\beta} = -R^\mu_{\alpha\beta\nu}. \tag{2.55}$$

Furthermore, it is easy to verify that the following identity is valid

$$R^\alpha_{\beta\gamma\delta} + R^\alpha_{\delta\beta\gamma} + R^\alpha_{\gamma\delta\beta} = 0. \tag{2.56}$$

In addition to the mixed curvature tensor $R^\mu_{\alpha\nu\beta}$, we can also use the covariant curvature tensor

$$R_{\alpha\beta\gamma\delta} = g_{\alpha\eta}R^\eta_{\beta\gamma\delta}. \tag{2.57}$$

From the transformation law for $R_{\alpha\beta\gamma\delta}$

$$R_{\alpha\beta\gamma\delta} = \frac{1}{2}\left(\frac{\partial^2 g_{\alpha\delta}}{\partial x^\beta \partial x^\gamma} + \frac{\partial^2 g_{\beta\gamma}}{\partial x^\alpha \partial x^\delta} - \frac{\partial^2 g_{\alpha\gamma}}{\partial x^\beta \partial x^\delta} - \frac{\partial^2 g_{\beta\delta}}{\partial x^\alpha \partial x^\gamma}\right)$$

$$+ g_{\mu\nu}\left(\Gamma^\mu_{\beta\gamma}\Gamma^\nu_{\alpha\delta} - \Gamma^\mu_{\beta\delta}\Gamma^\nu_{\alpha\gamma}\right). \tag{2.58}$$

It is not difficult to derive this transformation law, but it is very tedious; hence we simply give it here without derivation. From (2.58) we see the following symmetry properties:

$$R_{\alpha\beta\gamma\delta} = -R_{\beta\alpha\gamma\delta} = -R_{\alpha\beta\delta\gamma}, R_{\alpha\beta\gamma\delta} = R_{\gamma\delta\alpha\beta} \tag{2.59}$$

i.e., the tensor is antisymmetric in each of the index pairs α, β, and γ, δ and is symmetric under the interchange of the two pairs with each other. Thus, all components $R_{\alpha\beta\gamma\delta}$, in which $\alpha = \beta$ or $\gamma = \delta$ are zero. As for $R_{\alpha\beta\gamma\delta}$ we also have the identity

$$R_{\alpha\beta\gamma\delta} + R_{\alpha\delta\beta\gamma} + R_{\alpha\gamma\delta\beta} = 0. \tag{2.60}$$

A fourth-rank tensor has $4^4 = 256$ components. However, because of the above symmetries, the number of algebraically independent components of $R_{\alpha\beta\gamma\delta}$ is only 20.

By contracting the curvature tensor, we get the symmetric Ricci tensor:

$$R_{\sigma\mu} = R^\lambda_{\sigma\mu\lambda} = R_{\mu\sigma}. \tag{2.61}$$

According to (2.54), we have

$$R_{\sigma\mu} = \frac{\partial \Gamma^\lambda_{\sigma\mu}}{\partial x^\lambda} - \frac{\partial \Gamma^\lambda_{\sigma\lambda}}{\partial x^\mu} + \Gamma^\lambda_{\sigma\mu}\Gamma^\nu_{\lambda\nu} - \Gamma^\nu_{\sigma\lambda}\Gamma^\lambda_{\mu\nu}. \tag{2.62}$$

This tensor is clearly symmetric: $R_{\sigma\mu} = R_{\mu\sigma}$.

Finally, contracting $R_{\sigma\mu}$ we obtain

$$R = g^{\sigma\mu}R_{\sigma\mu} = g^{\sigma\lambda}g^{\mu\beta}R_{\sigma\mu\lambda\beta}, \tag{2.63}$$

which is called the *scalar curvature of the space*.

One of the most important tensors in the study of gravitation is the Einstein tensor, defined by

$$G_{\mu\nu} = R_{\mu\nu} - \frac{1}{2}g_{\mu\nu}R = G_{\nu\mu}. \tag{2.64}$$

The Einstein tensor $G_{\mu\nu}$ is purely geometric in character, being built up from $g_{\mu\nu}$ and their first and second derivatives. And it is linear in the second derivatives of $g_{\mu\nu}$. It can be shown that the covariant divergence of $G_{\mu\nu}$ vanishes identically (we leave it as a problem). This property will be used later to formulate the gravitational field equations.

There is apparent obscurity surrounding the physical meaning of the Riemann tensor; a simple example of calculating the curvature of space doesn't elucidate the physical meaning of the Riemann tensor. This may be helpful to visualize the Riemann tensor. Let us consider the two-dimensional space on the surface of a sphere of radius a. The metric of this two-dimensional space is

$$ds^2 = a^2(d\theta^2 + \sin^2\theta d\phi^2).$$

The covariant and contravariant metric tensors are

$$g_{\mu\nu} = \begin{pmatrix} a^2 & 0 \\ 0 & a^2\sin^2\theta \end{pmatrix}, \quad g^{\mu\nu} = \begin{pmatrix} 1/a^2 & 0 \\ 0 & 1/(a^2\sin^2\theta) \end{pmatrix}$$

and $g = a^4\sin^2\theta$.

The Christoffel symbols are given by (2.42). By direction calculation, we find that the only non-zero symbols are

$$\Gamma^\theta_{\phi\phi} = \frac{1}{2}g^{\theta\alpha}\left(\frac{\partial g_{\phi\alpha}}{\partial\phi} + \frac{\partial g_{\alpha\phi}}{\partial\phi} - \frac{\partial g_{\phi\phi}}{\partial\alpha}\right), \text{ and } \Gamma^\phi_{\theta\phi} = \frac{1}{2}g^{\phi\alpha}\left(\frac{\partial g_{\theta\alpha}}{\partial\phi} + \frac{\partial g_{\phi\alpha}}{\partial\theta} - \frac{\partial g_{\theta\phi}}{\partial\alpha}\right)$$

where α takes the values θ and ϕ; these become

$$\Gamma^\theta_{\phi\phi} = -g^{\theta\theta}\frac{\partial g_{\phi\phi}}{\partial\theta} = -\sin\theta\cos\theta, \quad \Gamma^\phi_{\theta\phi} = \frac{1}{2}g^{\phi\phi}\frac{\partial g_{\phi\phi}}{\partial\theta} = \cot\theta.$$

Let us first calculate the Ricci tensor (the contracted Riemann tensor) that is given by (2.62). The non-zero components are

$$R_{\theta\theta} = \Gamma^\phi_{\theta\phi}\Gamma^\phi_{\theta\phi} + \frac{\partial\cot\theta}{\partial\theta} = \cot^2\theta - \frac{1}{\sin^2\theta} = -1$$

and

$$R_{\phi\phi} = -\sin^2\theta.$$

The Riemann scalar curvature R of the space is given by

$$R = g^{\theta\theta}R_{\theta\theta} + g^{\phi\phi}R_{\phi\phi} = -2/a^2.$$

If the reader is familiar with the analytical geometry of surfaces, then recall that R here is equivalent to

$$R = \frac{2}{\rho_1\rho_2}$$

where ρ_1 and ρ_2 are the two principal radii of curvature of the surfaces. On the sphere these radii coincide and are equal to the spherical radius. The Riemann scalar curvature thus bears a simple relationship to the radius curvature of the two-dimensional space on the spherical surface.

Proceeding similarly, we find, from (2.57), the only independent component of the Riemann tensor

$$R_{\theta\phi\theta\phi} = g_{\theta\theta}R^\theta_{\phi\theta\phi} = g_{\theta\theta}\left(\frac{\partial\Gamma^\theta_{\phi\phi}}{\partial\theta} - \frac{\partial\Gamma^\theta_{\theta\phi}}{\partial\phi} + \Gamma^\theta_{\theta\alpha}\Gamma^\alpha_{\phi\phi} - \Gamma^\theta_{\phi\alpha}\Gamma^\alpha_{\theta\phi}\right) = a^2\sin^2\theta,$$

and by (2.59) we have other non-zero components

$$R_{\phi\theta\phi\theta} = -R_{\theta\phi\phi\theta} = -R_{\phi\theta\theta\phi} = R_{\theta\phi\theta\phi} = a^2\sin^2\theta.$$

2.8 The Condition for Flat Space

If space is flat, we may choose a Cartesian coordinate system. Then the $g_{\mu\nu}$ are all constant and all Christoffel symbols vanish; hence, from (2.54),

$$R^{\mu}_{\alpha\nu\beta} = 0$$

in this coordinates system. But if $R^{\mu}_{\alpha\nu\beta} = 0$ in one coordinate system, the components are zero in all coordinate systems. Hence if a space is flat, the Riemann curvature tensor must vanish, which is a necessary condition for a space to be flat. The converse is a sufficient condition: if $R^{\mu}_{\alpha\nu\beta} = 0$, the space is flat. We now proceed to prove it.

If $R^{\mu}_{\alpha\nu\beta} = 0$, and if we can find a coordinate system for which its metric tensor is constant, then the space is flat. To this end, let us take vector A_{μ} located at Point x and parallel displace it to Point $x + dx$, and then parallel displace it to Point $x + dx + \delta x$. If the Riemann curvature tensor $R^{\mu}_{\alpha\nu\beta}$ is zero, the result would be the same if we had displace A_{μ} from x to $x + \delta x$ and then to $x + \delta x + dx$. That is, the displacement is independent of the path. Thus we can displace the vector to a distant point and the result we get is independent of the path to the distant point. If we displace the vector A_{μ} at x to all points by parallel displacement, we will get a vector field that satisfies $A_{\mu;\nu} = 0$:

$$A_{\mu;\nu} = \frac{\partial A_{\mu}}{\partial x^{\nu}} - \Gamma^{\sigma}_{\mu\nu} A_{\sigma} = 0, \quad \text{or} \quad A_{\mu,\nu} = \Gamma^{\sigma}_{\mu\nu} A_{\sigma}.$$

If A_{μ} is the gradient of a scalar S, $A_{\mu} = \partial S/\partial x^{\mu} = S_{,\mu}$, then the above equation becomes

$$S_{,\mu\nu} = \Gamma^{\sigma}_{\mu\nu} S_{,\sigma}.$$

Because $\Gamma^{\sigma}_{\mu\nu} = \Gamma^{\sigma}_{\nu\mu}$, $S_{,\mu\nu} = S_{,\nu\mu}$ the above equations can be integrated.

Now let us take four independent scalars satisfying the last equations and let them to be the coordinates y^{α} of a new coordinate system. Thus, we have

$$y^{\alpha}_{,\mu\nu} = \Gamma^{\sigma}_{\mu\nu} y_{,\sigma},$$

where $y^{\alpha}_{,\mu\nu} = \partial^2 y^{\alpha}/\partial x^{\mu} \partial x^{\nu}$.

Let us go back to Eq. (10) that now takes the form

$$g_{\mu\lambda}(x) = g_{\alpha\beta}(y) \frac{\partial y^{\alpha}}{\partial x^{\mu}} \frac{\partial y^{\beta}}{\partial x^{\lambda}}.$$

Differentiation of this equation with respect to x^{ν} yields

$$\begin{aligned}
\frac{\partial g_{\mu\lambda}}{\partial x^{\nu}} &= \frac{\partial g_{\alpha\beta}(y)}{\partial x^{\nu}} \frac{\partial y^{\alpha}}{\partial x^{\mu}} \frac{\partial y^{\beta}}{\partial x^{\lambda}} + g_{\alpha\beta}(y) \left(\frac{\partial^2 y^{\alpha}}{\partial x^{\mu} \partial x^{\nu}} \frac{\partial y^{\beta}}{\partial x^{\lambda}} + \frac{\partial y^{\alpha}}{\partial x^{\mu}} \frac{\partial^2 y^{\beta}}{\partial x^{\lambda} \partial x^{\nu}} \right) \\
&= \frac{\partial g_{\alpha\beta}(y)}{\partial x^{\nu}} \frac{\partial y^{\alpha}}{\partial x^{\mu}} \frac{\partial y^{\beta}}{\partial x^{\lambda}} + g_{\alpha\beta}(y) \left(\Gamma^{\sigma}_{\mu\nu} y^{\alpha}_{,\sigma} y^{\beta}_{,\lambda} + y^{\alpha}_{,\mu} \Gamma^{\sigma}_{\lambda\nu} y^{\beta}_{,\sigma} \right) \\
&= \frac{\partial g_{\alpha\beta}(y)}{\partial x^{\nu}} \frac{\partial y^{\alpha}}{\partial x^{\mu}} \frac{\partial y^{\beta}}{\partial x^{\lambda}} + g_{\alpha\lambda}(x) \Gamma^{\sigma}_{\mu\nu} + g_{\mu\sigma}(x) \Gamma^{\sigma}_{\lambda\nu}.
\end{aligned}$$

The last two terms can be rewritten in terms of the Christoffel symbols of the first kind:

$$g_{\alpha\lambda}(x)\Gamma^\sigma_{\mu\nu} + g_{\mu\sigma}(x)\Gamma^\sigma_{\lambda\nu} = \Gamma_{\lambda,\mu\nu} + \Gamma_{\mu,\lambda\nu} = \partial g_{\mu\lambda}/\partial x^\nu = g_{\mu\lambda,\nu}$$

where the Christoffel symbols are given by (2.40). Combining this with the last equation, we obtain

$$\frac{\partial g_{\alpha\beta}(y)}{\partial x^\nu}\frac{\partial y^\alpha}{\partial x^\mu}\frac{\partial y^\beta}{\partial x^\lambda} = 0.$$

It follows that

$$\frac{\partial g_{\alpha\beta}(y)}{\partial x^\nu} = 0.$$

Thus, the metric tensor of the new system of coordinates is constant, and the Christroffel symbols all vanish identically. In other words, we have flat space.

2.9 Geodesic Deviation

We can get a good insight into the nature of a space just by examining the problem of geodesic deviation. For example, consider two nearby freely falling particles which travel on paths $x^\mu(\tau)$ and $x^\mu(\tau) + \delta x^\mu(\tau)$. The equations of motion are then given by

$$\frac{d^2 x^\mu}{d\tau^2} + \Gamma^\mu_{\nu\lambda}(x)\frac{dx^\nu}{d\tau}\frac{dx^\lambda}{d\tau} = 0 \tag{2.65}$$

and

$$\frac{d^2}{d\tau^2}[x^\mu + \delta x^\mu] + \Gamma^\mu_{\nu\lambda}(x + \delta x)\frac{d}{d\tau}[x^\nu + \delta x^\nu]\frac{d}{d\tau}[x^\lambda + \delta x^\lambda] = 0. \tag{2.66}$$

Evaluating the difference between these equations to first order in δx^μ gives

$$\frac{d^2 \delta x^\mu}{d\tau^2} + \frac{\partial \Gamma^\mu_{\nu\lambda}}{\partial x^\rho}\delta x^\rho \frac{dx^\nu}{d\tau}\frac{dx^\lambda}{d\tau} + 2\Gamma^\mu_{\nu\lambda}\frac{dx^\nu}{d\tau}\frac{dx^\rho}{d\tau} = 0 \tag{2.67}$$

or, in terms of covariant derivatives along the curves $x^\mu(\tau)$,

$$\frac{D^2}{D\tau^2}\delta x^\lambda = R^\lambda_{\nu\mu\rho}\delta x^\mu \frac{dx^\nu}{d\tau}\frac{dx^\rho}{d\tau}. \tag{2.68}$$

This (2.68) is called the *equation of geodesic deviation*.

Although a freely falling particle appears to be at rest in a coordinate system falling with the particle, a pair of nearby freely falling particles will exhibit a relative motion that can reveal the presence of a gravitational field to an observer who falls with the particles. The effect of the right-hand side of (2.68) becomes negligible when the separation between particles is much less than the characteristic dimensions of the field. This indicates clearly that the local inertial frames are only locally applicable; otherwise, the principle of equivalence will be violated.

2.10 Laws of Physics in Curved Spaces

The laws of physics must be valid in all coordinate systems. If they are expressed as tensor equations, whenever they involve the derivative of a field quantity, it must be a covariant derivative. Even if we are working with flat space (which means neglecting the gravitational field) and we are using curvilinear coordinates, we must write our equations in terms of covariant derivatives if we want them to hold in all coordinate systems.

As for the problem of generalizing a particular physics law from the flat Minkowski space to a general curved space, there is not a unique solution at all. In fact, the problem is highly complicated, since in general there will be an interaction between the space (the $g_{\mu\nu}$) and the physical phenomenon whose laws we are trying to formulate. But if the object under consideration does not appreciably influence the $g_{\mu\nu}$, that is, if the $g_{\mu\nu}$ are determined by objects much more massive than the object under consideration, we may then consider the $g_{\mu\nu}$ as given functions of the spacetime variables, $g_{\mu\nu}(x^\sigma)$. In this case the geometry is rigidly determined and the effect of the physical object under study on the geometrical structure may be neglected. Under these circumstances, we may take over the special relativistic laws by substituting

$$d \to D, \partial_\mu \to D_\mu, d\Omega \to \sqrt{-g}\,d\Omega$$

where

$$\partial_\mu A^\nu := \frac{\partial A^\nu}{\partial x^\mu}, D_\mu A = A^\mu{}_{;\nu}.$$

As an example, consider the motion of a free particle. Its time track in Minkowski space is characterized by the equations

$$d^2 x^i / ds^2 = 0, i = 0, 1, 2, 3. \tag{2.69}$$

These equations imply a straight line in the four-dimensional Minkowski space, which in turn corresponds to a uniform rectilinear motion in three-dimensional space. These equations can be derived from the stationary condition that the integral $\int ds$, taken along the motion between two points P and Q is stationary if one makes a small variation of the path keeping the end points fixed:

$$\delta \int_P^Q ds = 0 \tag{2.70}$$

where the variation vanishes at the end points P and Q.

If the particle is subject to the action of a gravitational field, its equation of motion is no longer a straight line, because the spacetime is curved. But the particle still follows a stationary trajectory, the geodesic. As shown above, the geodesic equation can be obtained from the same variational principle (2.70), provided that the Minkowski metric is replaced by the curved space metric $g_{\mu\nu}$ and $ds^2 = g_{\mu\nu} dx^\mu dx^\nu$. Instead of computing explicitly the variation in (2.70), we can simply obtain the geodesic equations as the covariant generalization of (2.69):

$$D^2 x^i / Ds^2 = 0,$$

which is equivalent to

$$d^2 x^i / ds^2 + \Gamma^\alpha_{\mu\nu} (dx^\mu / ds)(dx^\nu / ds) = 0,$$

the geodesic equations.

2.11 The Metric Tensor and the Classical Gravitational Potential

The presence of a gravitational field modifies the structure of spacetime. Any gravitational field is just a change in the metric of space-time, as determined by the metric tensor $g_{\mu\nu}$. Through the geodesic equations of motion, we can now provide the expressions governing the union of geometry and gravitation. To this end, let us compare the Newtonian equation of motion of a particle in a gravitational field and its geodesic equations of motion in a curved-space geometry:

$$d^2 x^\alpha / ds^2 + \partial\phi / \partial x^\alpha = 0 \tag{2.71}$$

$$d^2 x^\alpha / ds^2 + \Gamma^\alpha{}_{\mu\nu} (dx^\mu / ds)(dx^\nu / ds) = 0 \tag{2.72}$$

where ϕ is the Newtonian gravitational potential. These two equations have a fundamental similarity in that both are independent of the mass of the moving body under consideration. Thus, both equations satisfy the principle of equivalence. Now since the derivative $d^2 x^\alpha / ds^2$ is the four-acceleration of the particle, the quantity $-m\Gamma^\alpha{}_{\mu\nu} u^\mu u^\nu$ may be interpreted as the gravitational force, and then the components of the metric tensor $g_{\mu\nu}$ play the role of the Newtonian gravitational potential ϕ (as the Christoffel symbols are constructed from the derivatives of the $g_{\mu\nu}$). We must first show that this interpretation is consistent with the Newtonian equations of motion; namely, we must show that in the limit of ordinary velocities the geodesic equations reduce to the Newtonian equations. To see this, let the velocity $dx^\alpha / dt \ll c$. Then $ds^2 = g_{00} c^2 dt^2$, and $ds = \sqrt{g_{00}} c dt$, so that

$$d^2 x^\alpha / dt^2 + \Gamma^\alpha{}_{00} c^2 = 0$$

where

$$\Gamma^\alpha{}_{00} = \partial g_{00} / \partial x^\alpha.$$

From this we see that in this limit $g_{00} = K + 2\phi/c^2$. Since in flat space $g_{00} = 1$, we have $K = 1$ and

$$g_{00} = 1 + 2\phi/c^2 \tag{2.73}$$

This shows that the identification postulated above is plausible, i.e., the metric tensor $g_{\mu\nu}$ plays the role of Newtonian gravitational potential.

We should be careful to note that the physical content of the two equations, (2.71 and 2.72), are entirely different. In Newton's equation we have a field ϕ, which causes the motion. The particle is under a force and its velocity changes in time. In the geodesic equations, on the other hand, there is no physical agent such as ϕ. The particle follows a geodesic that is determined by the geometry of the space-time. This change in interpretation is actually a conceptual simplification, since inertia and gravitation are unified and the concept of external force is eliminated from the theory of gravitation.

2.12 Some Useful Calculation Tools

We conclude this chapter by giving a number of useful aids in manipulation of tensor quantities. First of all, it is very helpful to bear in mind that the covariant derivative of the metric tensor $g_{\mu\nu}$ is zero. That is, the metric tensor $g_{\mu\nu}$ may be considered as a constant during covariant differentiation.

We now derive an expression for the contracted Christoffel symbol $\Gamma^{\mu}_{\alpha\mu}$ that will be very useful later on. From (2.39) we have

$$\Gamma^{\mu}_{\alpha\mu} = \frac{1}{2} g^{\mu\nu} \left(\frac{\partial g_{\nu\alpha}}{\partial x^{\mu}} + \frac{\partial g_{\nu\mu}}{\partial x^{\alpha}} - \frac{\partial g_{\alpha\mu}}{\partial x^{\nu}} \right).$$

Changing the positions of the dummy indexes μ and ν in the first term and remembering $g_{\mu\nu} = g_{\nu\mu}$, we see that the first and third terms then cancel each other, so that

$$\Gamma^{\mu}_{\alpha\mu} = \frac{1}{2} g^{\mu\nu} \frac{\partial g_{\mu\nu}}{\partial x^{\alpha}}, \tag{2.74}$$

which can be simplified. To do this we calculate the differential dg of the determinant g made up from the components of the metric tensor $g_{\mu\nu}$; dg can be obtained by taking the differential of each component of the tensor $g_{\mu\nu}$ and multiplying it by its coefficient in the determinant, i.e., by the corresponding minor:

$$dg = dg_{\mu\nu} M^{\mu\nu}$$

where $M^{\mu\nu}$ is the minor of the component $g_{\mu\nu}$. Now,

$$g^{\mu\nu} = \frac{M^{\mu\nu}}{g}, \qquad M^{\mu\nu} = g^{\mu\nu} g$$

Thus,

$$dg = g g^{\mu\nu} dg_{\mu\nu} = -g g_{\mu\nu} dg^{\mu\nu}$$

The expression on the far right of the above equation follows from

$$d(g_{\mu\nu} g^{\mu\nu}) = d(\delta^{\mu}_{\mu}) = d(4) = 0.$$

We then have

$$\frac{\partial g}{\partial x^{\alpha}} = g g^{\mu\nu} \frac{\partial g_{\mu\nu}}{\partial x^{\alpha}} = -g g_{\mu\nu} \frac{\partial g^{\mu\nu}}{\partial x^{\alpha}}. \tag{2.75}$$

The use of (2.75) enables us to write (2.74) in the form

$$\Gamma^{\mu}_{\alpha\mu} = \frac{1}{2} g^{\mu\nu} \frac{\partial g_{\mu\nu}}{\partial x^{\alpha}} = \frac{1}{2g} \frac{\partial g}{\partial x^{\alpha}} = \frac{\partial \ln \sqrt{-g}}{\partial x^{\alpha}}. \tag{2.76}$$

This expression is very useful. First consider the covariant divergence $A^{\mu}_{;\mu}$:

$$A^{\mu}_{;\mu} = \frac{\partial A^{\mu}}{\partial x^{\mu}} + \Gamma^{\mu}_{\alpha\mu} A^{\alpha}.$$

Substituting the expression (2.76) for $\Gamma^{\mu}_{\alpha\mu}$, we obtain

$$A^{\mu}_{;\mu} = \frac{\partial A^{\mu}}{\partial x^{\mu}} + A^{\alpha} \frac{\partial \ln \sqrt{-g}}{\partial x^{\alpha}} = \frac{1}{\sqrt{-g}} \frac{\partial \left(\sqrt{-g} A^{\mu}\right)}{\partial x^{\mu}}. \tag{2.77}$$

We now consider the covariant divergence of a contravariant tensor of the second rank $T^{\mu\nu}_{;\alpha}$. From (2.27), we have

$$T^{\mu\nu}_{;\alpha} = \frac{\partial T^{\mu\nu}}{\partial x^{\nu}} + \Gamma^{\mu}_{\beta\nu} T^{\beta\nu} + \Gamma^{\mu}_{\beta\alpha} T^{\mu\beta}.$$

Changing the positions of the dummy indexes μ and β in the third term on the right-hand side, we obtain

$$T^{\mu\nu}_{;\nu} = \frac{\partial T^{\mu\nu}}{\partial x^{\nu}} + \Gamma^{\mu}_{\beta\nu} T^{\beta\nu} + \Gamma^{\beta}_{\nu\beta} T^{\mu\nu}.$$

Substituting the expression (2.76) for $\Gamma^{\beta}_{\nu\beta}$, we obtain

$$T^{\mu\nu}_{;\nu} = \frac{\partial T^{\mu\nu}}{\partial x^{\nu}} + \Gamma^{\mu}_{\beta\nu} T^{\beta\nu} + \frac{\partial \ln \sqrt{-g}}{\partial x^{\nu}} T^{\mu\nu}$$

or

$$T^{\mu\nu}_{;\nu} = \frac{1}{\sqrt{-g}} \frac{\partial}{\partial x^{\nu}} (\sqrt{-g} T^{\mu\nu}) + \Gamma^{\mu}_{\beta\nu} T^{\beta\nu}. \tag{2.78}$$

Similarly, for a mixed tensor, (2.28) leads to

$$T^{\beta}_{\alpha;\beta} = \frac{1}{\sqrt{-g}} \frac{\partial}{\partial x^{\beta}} \left(T^{\beta}_{\alpha} \sqrt{-g}\right) - \Gamma^{\beta}_{\alpha\sigma} T^{\alpha}_{\beta}. \tag{2.79}$$

For an antisymmetric tensor $F^{\beta\nu} = -F^{\nu\beta}$, then

$$\Gamma^{\mu}_{\beta\nu} F^{\beta\nu} = \Gamma^{\mu}_{\nu\beta} F^{\nu\beta} = -\Gamma^{\mu}_{\beta\nu} F^{\beta\mu}. \tag{2.80}$$

The expression in the middle is obtained by interchanging the dummy indexes β and ν, and the expression on the far right follows from the $F^{\beta\nu} = -F^{\nu\beta}$, and $\Gamma^{\mu}_{\nu\beta} = \Gamma^{\mu}_{\beta\nu}$. From (2.80) it follows that

$$\Gamma^{\mu}_{\beta\nu} F^{\beta\mu} = 0.$$

Thus, for an antisymmetric tensor $F^{\alpha\beta}$, the last term of (2.78) vanishes and the covariant divergence is

$$F^{\alpha\beta}_{;\beta} = \frac{1}{\sqrt{-g}} \frac{\partial}{\partial x^\beta} (F^{\alpha\beta} \sqrt{-g}). \tag{2.81}$$

For a symmetric tensor $S^{\alpha\beta}$, rearrangement of the last term of (2.79) gives

$$S^\beta_{\alpha;\beta} = \frac{1}{\sqrt{-g}} \frac{\partial}{\partial x^\beta} (S^\beta_\alpha \sqrt{-g}) - \frac{1}{2} \frac{\partial g_{\beta\nu}}{\partial x^\alpha} S^{\beta\nu}. \tag{2.82}$$

In Cartesian coordinates, the curl

$$\frac{\partial A_\mu}{\partial x^\nu} - \frac{\partial A_\nu}{\partial x^\mu}$$

is an antisymmetric tensor. In curvilinear coordinates this tensor is $A_{\mu;\nu} - A_{\nu;\mu}$:

$$A_{\mu;\nu} - A_{\nu;\mu} = A_{\mu,\nu} - \Gamma^\rho_{\mu\nu} A_\rho - (A_{\nu,\mu} - \Gamma^\rho_{\nu\mu} A_\rho)$$

Since $\Gamma^\rho_{\mu\nu} = \Gamma^\rho_{\nu\mu}$, we have

$$A_{\mu;\nu} - A_{\nu;\mu} = \frac{\partial A_\mu}{\partial x^\nu} - \frac{\partial A_\nu}{\partial x^\mu}. \tag{2.83}$$

This result may be stated: covariant curl equals ordinary curl, but it holds only for a covariant vector. For a contravariant vector we could not form the curl because the suffixes would not balance.

Finally, we transform to curvilinear coordinates the sum

$$\frac{\partial^2 \varphi}{\partial x_\alpha \partial x^\alpha}$$

of the second derivatives of a scalar φ. In curvilinear coordinates this sum goes over to $\varphi^{;\alpha}_{;\alpha}$. But covariant differentiation of a scalar reduces to ordinary differentiation:

$$\varphi_{;\alpha} = \partial \varphi / \partial x^\alpha.$$

Raising the index α, we have

$$\varphi^{;\alpha} = g^{\alpha\beta} \partial \varphi / \partial x^\beta,$$

which is a contravariant tensor of rank one. Using formula (2.77), we find

$$\varphi^{;\alpha}_{;\alpha} = \frac{1}{\sqrt{-g}} \frac{\partial}{\partial x^\alpha} \left(\sqrt{-g}\, g^{\alpha\beta} \frac{\partial \varphi}{\partial x^\beta} \right). \tag{2.84}$$

In view of Eq. (77), Gauss' theorem for transformation of the integral of a vector over a hypersurface into an integral over a four-volume can now be written as

$$\oint A^\alpha \sqrt{-g}\, dS_\alpha = \int A^\alpha_{;\alpha} \sqrt{-g}\, d\Omega. \tag{2.85}$$

2.13 Problems

2.1. Write the terms in each of the following indicated sums

$$a \cdot a_{jk}x^k \quad b \cdot A_{pq}B^{qr} \quad c.\bar{g}_{rs} = g_{jk}\frac{\partial x^j}{\partial \bar{x}^r}\frac{\partial x^k}{\partial \bar{x}^s}$$

2.2. Write the transformation law for the following tensors: $a \cdot A^i_{\ jk} \quad b \cdot B^{pq}_{\ \ ijk}$

2.3. A quantity $A(j, k, m, n)$, which is a function of the coordinates x^i, transforms to another coordinate system $\bar{x}^i$ according to the rule

$$\bar{A}(p, q, r, s) = \frac{\partial x^j}{\partial \bar{x}^p}\frac{\partial \bar{x}^q}{\partial x^k}\frac{\partial \bar{x}^r}{\partial x^m}\frac{\partial \bar{x}^s}{\partial x^n}A(j, k, m, n).$$

Is the quantity a tensor? If so, write the tensor in suitable notation and give the covariant and contravariant rank.

2.4. Show that the property of symmetry (or antisymmetry) with respect to indexes of a tensor is invariant under coordinate transformation.

2.5. A covariant tensor has components $xy, 2y - z^2, xz$ in rectangular coordinates; find its covariant components in spherical coordinates.

2.6. Prove that the contraction of the outer product of the two tensors A^p and B_p is a scalar.

2.7. Determine the Christoffel symbols of the second kind in rectangular and cylindrical coordinates.

2.8. The line element on the surface of a sphere of radius a in Euclidean space is given by $ds^2 = a^2(d\theta^2 + \sin^2\theta d\phi^2)$. For this space calculate $\Gamma^i_{\ kl}, i, k, l = 1, 2$ (with $\theta = x^1, \phi = x^2$).

2.9. Find the covariant derivative of $A^i_{\ j}B^{km}_{\ \ n}$ with respect to x^q.

2.10. Prove that

$$\nabla \cdot A^p = \frac{1}{\sqrt{-g}}\frac{\partial}{\partial x^k}\left(\sqrt{-g}A^k\right) \quad \text{and} \quad \nabla^2\phi = \frac{1}{\sqrt{-g}}\frac{\partial}{\partial x^k}\left(\sqrt{-g}g^{kr}\frac{\partial\phi}{\partial x^r}\right).$$

2.11. Determine the force acting on a particle in a constant gravitational field.

2.12. Prove that the covariant divergence of the Einstein tensor vanishes.

2.13. The distance s between two points on a curve $x^\mu = x^\mu(\lambda)$ is given by

$$s = \int ds = \int_1^2 \sqrt{g_{\alpha\beta}\frac{dx^\alpha}{d\lambda}\frac{dx^\beta}{d\lambda}}.$$

Show that the necessary condition that s be an extremum is that

$$\frac{\partial L}{\partial x^\nu} - \frac{d}{d\lambda}\frac{\partial L}{\partial \dot{x}^\nu} = 0$$

where

$$L = g_{\alpha\beta}\frac{dx^\alpha}{d\lambda}\frac{dx^\beta}{d\lambda}, \text{ and } \dot{x}^\nu = \frac{dx^\nu}{d\lambda}.$$

2.14. Show that great circles drawn on the surface of a sphere are geodesics.

2.15. Show that, with (2.43), in a Euclidean space geodesics are straight lines.

References

Dirac PAM (1975) *General Theory of Relativity*. (John Wiley & Sons, New York)
Landau LD, Lifshitz EM (1975) *The Classical Theory of Fields*. (Pergamon Press, Oxford UK)

Chapter 3
Einstein's Law of Gravitation

3.1 Introduction (Summary of General Principles)

We now summarize the concepts and ideas discussed so far as a series of postulates. They will indicate the correct approach for establishing the new laws of gravitation.

(1) Under a general coordinate transformation, the laws of physics remain covariant (i.e., form invariant). This requirement is known as the principle of general covariance.
(2) When gravitational fields are present, space-time is curved and endowed with a metric of the form

$$ds^2 = g_{\mu\nu}dx^\mu dx^\nu \quad (\mu, \nu = 0, 1, 2, 3).$$

The metric tensor $g_{\mu\nu}$ gives functions of the space-time variables x^0, x^1, x^2, and x^3; and there are 10 of them. If we include the effect of coordinate freedom there are only 6 independent $g_{\mu\nu}$.
(3) The metric tensor $g_{\mu\nu}$ can be interpreted as the generalization of the gravitational field Φ. This is an indication of the union of gravitation and geometry.
(4) According to Einstein, the curvature of space-time is governed by the masses embedded in it.
(5) In the nonrelativistic limit, Newtonian dynamics and Newtonian gravitational theory are both valid. We can call this the *correspondence principle*.

According to (1), the new laws of gravitation must be expressed in tensor form. While (2) and (3) suggest that the new laws contain at least six quantities that are related to $g_{\mu\nu}$, (4) says that the energy-momentum tensor must be related to the curvature of space-time. Finally, (5) says that in the nonrelativistic limit, our new laws of gravitation reduce to Poisson's equation.

3.2 A Heuristic Derivation of Einstein's Equations

It is not clear how one can formulate precisely the laws of gravitation starting from such extremely general postulates. It is Einstein's triumph and an example of the power of speculative thought that he was able to formulate the laws of gravitation from such general principles. We follow Chandrasekhar's approach in formulating the laws of gravitation, namely, we ask three basic questions, and answer them first in terms of Newtonian theory, then via their relativistic generalizations. The three basic questions are the following:

(1) When can we say that there is no gravitational field present?
(2) What are the equations that determine the gravitational field in empty space, outside material bodies?
(3) What are the equations in regions of space where matter is present?

In the Newtonian theory, gravitational field is to be described in terms of a scalar potential function Φ, and the equations that determine Φ are

(*i*)

$$\Phi = 0. \tag{3.1}$$

and when there is no gravitation;

(*ii*)

$$\nabla^2 \Phi = 0 \text{ (Laplace equation)} \tag{3.2}$$

in empty space (no matter present and no physical fields except a gravitational field); and finally

(*iii*)

$$\nabla^2 \Phi = 4\pi G\rho \text{ (Poisson's equation)} \tag{3.3}$$

in regions of space where matter is present and the material density is ρ. These equations are supplemented by the standard equations of motion.

We first seek relativistic generalizations of (i) and (ii). The generalization of (iii) will be addressed later.

3.2.1 Vacuum Field Equations

How can we generalize the condition $\Phi = 0$? We start with Newton's first law of motion that states that in the absence of a gravitational field, a free particle can experience no acceleration, i.e., in a Cartesian frame:

$$d^2x^i/ds^2 = 0. \tag{3.4}$$

If we should describe the motion in curvilinear coordinates, the particle will be subjected to inertial accelerations and the equation of motion of the particle is a geodesic equation (see Eq. 2.44):

$$d^2x^\nu/ds^2 + \Gamma^\nu_{\mu\sigma}(dx^\mu/ds)(dx^\sigma/ds) = 0. \tag{3.5}$$

We may therefore conclude that no gravitational field is present if a coordinate transformation is possible so that in the new coordinate system the equations of motion reduce to (3.4); in other words, in the new coordinate system the Christoffel symbols $\Gamma^\nu_{\mu\sigma}$ all vanish identically. As shown in Chapter 2 (Section 2.8), such a coordinate transformation is possible, and the metric tensor of the new system of coordinates is constant. Then the Christoffel symbols and so the Riemann tensor $R^\alpha{}_{\beta\gamma\delta}$ all vanish identically. We may now state that the condition for the absence of any gravitational field is the vanishing of the Riemann curvature tensor:

$$R^\alpha{}_{\beta\gamma\delta} = 0. \tag{3.6}$$

This corresponds to the requirement $\Phi = 0$ in the Newtonian theory.

We now need a generalization of (3.6), in the sense that in the Newtonian theory $\nabla^2\Phi = 0$ is a generalization of $\Phi = 0$. The equation $\nabla^2\Phi = 0$ is satisfied by $\Phi = 0$, but it allows nonvanishing solutions as well. The simplest such generalization of (3.6) appears to be

$$R^\mu{}_{\nu\rho\mu} = R_{\nu\rho} = 0. \tag{3.7}$$

Eq. (3.7) is a second-order partial differential equation as Laplace's equation is; it also provides six equations for the six independent coefficients of the metric tensor. We can support this generalization with the following argument. As shown in Chapter 2 (Section 2.11), to first order in $1/c^2$, we have

$$g_{00} = 1 - 2\Phi/c^2.$$

This equation implies that

$$\lim_{c\to\infty} \frac{1}{2}c^2\nabla^2 g_{00} = -\nabla^2\Phi.$$

Further, defining the d'Alembertian

$$\Box^2 = (1/c^2)\partial^2/\partial t^2 - \partial^2/\partial x^2 - \partial^2/\partial y^2 - \partial^2/\partial z^2$$

we see that

$$\lim_{c\to\infty} \frac{1}{2}\Box^2 g_{00} = \nabla^2\Phi.$$

This equation strongly suggests that the generalization of $\nabla^2\Phi = 0$ to curved spacetime should be a second-order tensor that (1) contains the metric tensor and its derivatives up to at most second order, and (2) is linear in the derivatives of the second order.

In fact, Einstein proposed to use (3.7) as a law of gravitation in empty space; we shall simply refer it as the vacuum field equations. Empty or vacuum here means that there is no matter present and no physical fields except the gravitational field. The gravitational field does not disturb the emptiness. Other fields do.

The vacuum field equations (3.7) lead to $R = 0$, and hence

$$R^{\mu\nu} - \frac{1}{2}g^{\mu\nu}R = 0. \tag{3.8}$$

We may either use (3.7) or (3.8) as the basic equations for empty space.

Now, instead of one field equation in Newtonian theory, there are 10 in the Einstein theory. They describe not only the gravitational field but also the system of coordinates. The gravitational field and the system of coordinates are inextricably bound up in the Einstein theory, and one cannot describe the one without the other.

The vacuum field equations are like the usual field equations of physics in that they are of the second order, because second derivatives appear in (3.7), as the Christoffel symbols involve first derivatives. But the vacuum field equations are unlike the usual field equations in that they are not linear. The nonlinearity means that the equations are complicated and it is difficult to get accurate solutions.

3.2.2 Field Equations Where Matter is Present in Space

The final step in generalizing the Newtonian theory is to write the analog of Poisson's equation (3.3):

$$\nabla^2 \Phi = 4\pi G\rho$$

where ρ is the density of matter, and G the Newton gravitational constant. The left side of (3.3) may be made Lorentz-invariant by writing

$$\Box^2 \Phi = 4\pi G\rho. \tag{3.9}$$

Mass, however, is not an invariant, so the right side of (3.9) is not a component of four-vector. We now recall the conservation law of four-momentum in special relativity

$$dP_\mu = 0, \quad P_\mu = \sum_\mu p_\mu$$

where P_μ is the four-momentum of the system and p_μ is the four-momentum of individual particle. (If the reader needs review of special relativity, Appendix II at the end of the book is there for this purpose. There is one warning: here we use covariant four-momentum P_μ instead contravariant one P^μ.) If matter-energy is continuously distributed in space, the formulas will take different forms. The continuous distributions may be considered as a limiting case of discrete particles, where in each volume element the number of point particles tends to infinity while the mass of each tends to zero. Conversely, a point particle may be looked at as a limiting case of a continuous distribution where at some points (the locations of the particles) the densities become infinitely large. With this understanding, we now consider, for example, the energy density $\varepsilon\{x\}$. We usually write the total energy as $E = \int \varepsilon(x)dV$, where dV is the volume element and it is the zeroth component of a four-dimensional surface element $dS^\nu : dS^0 = dV = dxdydz$. The energy is the zeroth component of the four-momentum vector. This suggest to us that the general form of the expression for the four-momentum vector is of the form

$$P_\mu = \int T_{\mu\nu} dS^\nu \tag{3.10}$$

which reduces to $E = \int T_{00} dS^0$ when the volume element and the matter in it are at rest. $T_{\mu\nu}$ is called the energy-momentum tensor (or the stress-energy tensor). The conservation of four-momentum (conservation of energy and momentum) require

$$dP_\mu = d \int T_{\mu\nu} dS^\nu = 0.$$

This can be written as (see Appendix II)

$$dP_\mu = \int \partial^\nu T_{\mu\nu} d\Omega = 0$$

from which

$$\frac{\partial T_\mu{}^\nu}{\partial x^\nu} = 0 \tag{3.11}$$

In general relativity we expect (3.11) is replaced by the covariant generalization:

$$T^{\mu\nu}{}_{;\nu} = 0 \tag{3.12}$$

where the semicolon denotes covariant differentiation. Next, we notice that the covariant divergence of the Ricci tensor $R_{\mu\nu}$ does not vanish, but the covariant divergence of the Einstein tensor

$$G^{\mu\nu} = R^{\mu\nu} - \frac{1}{2} g^{\mu\nu} R \tag{3.13}$$

does vanish. We therefore write

$$G^{\mu\nu} = \kappa T^{\mu\nu} \tag{3.14}$$

as the required law of gravitation in regions of space where matter is present. For the present the coupling constant κ is unspecified.

Tracing the Einstein tensor we obtain

$$G_\mu^\mu = g^{\mu\nu} G_{\mu\nu} = R - 2R = -R = \kappa T$$

where $T = T_\mu{}^\mu$. Therefore the generalized field equations (3.14) can also be rewritten

$$R_{\mu\nu} = \kappa \left(T_{\mu\nu} - \frac{1}{2} g_{\mu\nu} T \right). \tag{3.15}$$

It remains to specify the coupling constant κ in (3.14) and (3.15). When κ is appropriately chosen, we require that in the limit $c \to \infty$, Einstein's equations (3.15) must lead to Poisson's equation. To this aim, let us consider the motion of a particle that follows a geodesic equation:

$$\frac{d^2 x^\mu}{ds^2} + \Gamma_{\alpha\beta}^\mu \frac{dx^\alpha}{ds} \frac{dx^\beta}{ds} = 0. \tag{3.16}$$

In the nonrelativistic limit, neglecting terms of second order in the velocity, we have $ds^2 = c^2 dt^2 = (dx^0)^2$, and (3.16) reduces to

$$\frac{d^2 x^\mu}{dt^2} = -\Gamma^\mu_{00} \left(\frac{dx^0}{dt} \right)^2 = -c^2 \Gamma^\mu_{00}. \tag{3.17}$$

In the limit of weak gravitational field, we can put

$$g_{\mu\nu} = \eta_{\mu\nu} + h_{\mu\nu} \tag{3.18}$$

with

$$\eta_{00} = 1, \eta_{0i} = 0, \eta_{ij} = -\delta_{ij}, \text{ and } |h_{\mu\nu}| \ll 1 \tag{3.19}$$

and we can neglect terms of order h^2 and higher. We obtain then for a static field (so $\partial g_{\mu\nu}/\partial x^0 = 0$), from the definition of the Christoffel symbols

$$\Gamma^k_{00} = \frac{1}{2} g^{km} \left(\frac{\partial g_{00}}{\partial x^m} \right) = \frac{1}{2} \frac{\partial g_{00}}{\partial x^k} \tag{3.20}$$

where the Latin indexes take only three spatial values: 1,2, and 3. The geodesic equation now becomes

$$\frac{d^2 \vec{x}}{dt^2} = -\frac{c^2}{2} \nabla h_{00} \tag{3.21}$$

and coincides with Newton's equation of motion

$$\frac{d^2 \vec{x}}{dt^2} = -\nabla \Phi \tag{3.22}$$

provided that

$$h_{00} = \frac{2}{c^2} \Phi. \tag{3.23}$$

In the nonrelativistic limit, the leading term in the energy momentum tensor is $T_{00} = \rho c^2$, and the trace is then

$$T = g^{\mu\nu} T_{\mu\nu} \cong g^{00} T_{00} \cong \eta^{00} T_{00} = \rho c^2. \tag{3.24}$$

From (3.15) we then have

$$R_{00} = \frac{1}{2} \kappa \rho c^2. \tag{3.25}$$

Neglecting temporal derivatives, and Γ^2 terms, (2.62) gives in this limit

$$R_{00} = \frac{\partial \Gamma^k_{00}}{\partial x^k} = -\frac{1}{2} \nabla^2 h_{00} = -\frac{1}{c^2} \nabla^2 \Phi \tag{3.26}$$

Combining (3.25) and (3.26) we finally obtain

$$\nabla^2 \Phi = -\frac{1}{2} \kappa \rho c^2, \tag{3.27}$$

which is identical to the Poisson's equation if we set

$$\kappa = -\frac{8\pi G}{c^4}.$$ (3.28)

With this determination of κ, (3.15) becomes

$$G^{\mu\nu} \equiv R^{\mu\nu} - \frac{1}{2}g^{\mu\nu}R = -\frac{8\pi G}{c^4}T^{\mu\nu}$$ (3.29)

and the "derivation" of Einstein's field equations (Einstein's laws of gravitation) is completed.

Einstein's field equations (3.29) need not be supplemented by any statement concerning equations of motion that are given by

$$T^{\mu\nu}{}_{;\nu} = 0$$ (3.30)

Since the Einstein tensor $G^{\mu\nu}$ has zero covariant divergence, Einstein's field equations necessarily require that the covariant divergence of $T^{\mu\nu}$ vanish. Thus, Einstein's field equations contain the equations of motion. This is in contrast to the situation in electrodynamics, where Maxwell's equations lack the corresponding equation of motion.

3.3 Energy-Momentum Tensor

How is the energy momentum tensor to be specified in general relativity for a given physical system? The prescription for this is the following: *The required expression for $T^{\mu\nu}$ is the covariant generalization of the one valid in the special relativity.* For example, for a perfect fluid described in terms of an energy density ε and a scalar pressure p, the correct expression for its energy momentum tensor is

$$T^{\mu\nu} = (\varepsilon + p)u^\mu u^\nu - g^{\mu\nu}p, \left(u^\nu = \frac{dx^\nu}{ds}\right)$$ (3.31)

as this is the correct expression in special relativity. The justification for this prescription is that since we can always set up a Galilean coordinate frame *locally* and since the Einstein tensor is invariant to general coordinate transformations (i.e., covariant) the expression for $T^{\mu\nu}$ in a frame which is not locally Galilean should be obtainable by subjecting $T^{\mu\nu}$ (valid in the local Galilean frame) to the same tensor transformations.

We now digress for a moment to check the correctness of the expression (3.31). Consider a fluid consisting of a collection of particles with random motion. The particles collide, change directions, and generate pressures. We now evaluate the components of $T^{\mu\nu}$ in the frame in which the fluid as a whole is at rest (i.e., in the proper frame of the fluid). The four-momentum vector for a typical particle is given by

$$p^0 = \frac{mc^2}{\left(1 - v^2/c^2\right)^{1/2}}, \quad p^j = \frac{m\vec{v}}{\left(1 - v^2/c^2\right)^{1/2}} \quad (j = 1, 2, 3)$$

Then,

$$T^{00} = \sum mc^2 \left(1 - v^2/c^2\right)^{-1/2} \approx \sum mc^2 \left(1 + v^2/2c^2\right) = \rho c^2$$

$$T^{11} = T^{22} = T^{33} = \frac{1}{3} \sum mv^2 \left(1 - v^2/c^2\right)^{-1/2} \approx p,$$

where ρ and p are the density and pressure of the fluid. In terms of four-velocity u^μ, the energy-momentum tensor becomes

$$T^{\mu\nu} = \left(p + \rho c^2\right) u^\mu u^\nu - p\eta^{\mu\nu}$$

and its generally covariant form is

$$T^{\mu\nu} = \left(p + \rho c^2\right) u^\mu u^\nu - pg^{\mu\nu}.$$

3.4 Gravitational Radiation

The Theory of General Relativity is a relativistic theory of gravitation and predicts the existence of gravitational radiation. Unfortunately, it is extremely difficult to explore gravitational radiation from the full Einstein's field equations, not only in mathematics but even conceptually. This could be explained qualitatively as follows:

In electromagnetic radiation, it is the electric and magnetic fields that propagate as waves with the speed of light. What propagates in gravitational radiation? The answer unfortunately is not as clear as the electromagnetic waves. The gravitational effects in relativity are intimately related to the geometric structure of space-time. Hence we expect the structural changes in space-time to propagate as "gravitational waves." In practice, it is very difficult to single out any particular quantity that relates to such changes of space-time structure and that we can claim to be propagating as waves.

The difficulty lies partly in the coordinate description of space and time. Einstein's field equations have the beautiful property that they have the same formal structure, whatever the coordinate frame of reference used. But every observer uses a coordinate system to describe the geometric properties of space-time. The above-mentioned property gets in the way of deciding whether a particular solution does represent a gravitational wave or it is a result of the choice of a particular frame of reference. When gravitational fields are strong and the geometric properties of space-time are very different from Euclid's, the problem of interpreting a disturbance as a gravitational wave becomes very difficult. But in the case of weak gravitational fields it is simpler to identify certain disturbances as gravitational waves. For examples, massive bodies undergoing acceleration and two stars going around each other emit gravitational waves.

It is best to regard the weak-field solutions not as approximate solutions to the full equations, but as solutions to give an idea of the behavior expected in the full theory. We now outline the weak-field gravitational radiation, and refer the reader interested in the detailed development to the book by J. Weber listed as reference at the end of this chapter.

The weak-field wave solutions are obtained by supposing that the space is almost flat and an almost-Lorentz metric is appropriate. Then we can write the metric tensor as a Lorentz metric plus a small quantity of the first order:

$$g_{\mu\nu} = \delta_{\mu\nu} + h_{\mu\nu}. \tag{3.32}$$

We next define an additional quantity Φ^ν_μ:

$$\Phi^\nu_\mu = h^\nu_\mu - \frac{1}{2}\delta^\nu_\mu h^\alpha_\alpha \tag{3.33}$$

where h^α_α is the trace of h. It is possible to show that the quantity Φ^ν_μ satisfies a familiar wave equation:

$$\Box^2 \Phi^\nu_\mu = -\frac{16\pi G}{c^4}\tau^\nu_\mu \tag{3.34}$$

provided that we impose the subsidiary condition

$$\Phi^\nu_{\mu,\nu} = 0. \tag{3.35}$$

The Φ^ν_μ is the d'Alembertian of Φ^ν_μ, τ^ν_μ is the lowest order part of the energy-momentum tensor T^ν_μ.

These are the equations with which Einstein dealt. They are formally the same as those of electrodynamics. So we expect the stress-energy tensor to play the same role in gravitation theory as the four-current does in electromagnetic theory. It should be the source of the gravitational field, and hence the source of gravitational waves.

In 1939, Pauli and Fierz obtained a similar set of equations from quite different considerations. They were investigating the relativistic wave equations for particles of spin higher than 1/2, and they discovered that the appropriate relativistic wave equations for particles of spin 2 and rest-mass zero were the following:

$$\Box^2 \Phi^\nu_\mu = 0, \quad \Phi^\nu_{\mu,\nu} = 0. \tag{3.36}$$

These are the same as (3.34) and (3.35) for the vacuum case. This coincidence is hardly surprising. For spin-2 particles, a 10-component wave function is needed, five components for the spin and a doubling for the positive and negative energies. A second-rank symmetric tensor has 10 independent components, and so is a suitable representation of a 10-component wave function.

Since, for the vacuum case, the two sets of equations [(3.34), (3.35), and (3.36)] are formally the same, it follows that the particles of the gravitational field, the gravitons, will have spin 2. Since the gravitational field has infinite range, it follows also that the rest mass of the graviton is zero.

3.5 Problems

3.1. Prove the result shown in (3.31).

3.2. Given the space-time geometry of the following, find the geodesic equations of motion.

$$ds^2 = e^{-2\phi} dt^2 - e^{2\phi} \left(dx^2 + dy^2 + dz^2 \right).$$

3.3. If space-time has the metric

$$ds^2 = e^{\lambda} \left(dr^2 + dz^2 \right) + r^2 e^{-\rho} d\phi^2 - e^{\rho} dt^2$$

where λ, ρ are functions of r and z only, show that the field equations in empty space $R = 0$ require that following are satisfied:

$$\lambda_1 + \rho_1 = \frac{1}{2} r \left(\rho_1^2 - \rho_2^2 \right)$$

$$\lambda_2 + \rho_2 = r \rho_1 \rho_2$$

$$\rho_{11} + \rho_{22} + \frac{1}{r} \rho_1 = 0$$

$$\lambda_{11} + \lambda_{22} + \rho_{11} + \rho_{22} + \frac{1}{2} \left(\rho_1^2 + \rho_2^2 \right) = 0$$

where subscripts 1 and 2 denote partial differentiations with respect to r and z respectively.

3.4. If space-time has the metric

$$ds^2 = e^{2kx} \left(dx^2 + dy^2 + dz^2 - dt^2 \right)$$

where k is constant, and $v^2 = \dot{x}^2 + \dot{y}^2 + \dot{z}^2$, dots denoting differentiations with respect to t, show that for a freely falling body

$$1 - v^2 = \left(1 - V^2 \right) e^{2kx}$$

where $v = V$ at $x = 0$.

References

Chandrasekar S (Feb. 1972) On the Derivation of Einstein's Field Equations. Am J Phys 40 (2): 224–34

Landau LD, Lifshitz EM (1975) *The Classical Theory of Fields.* (Pergamon Press, Oxford, England)

Weber J (1961) *General Relativity and Gravitational Waves.* (Interscience Publishers Inc., New York)

Chapter 4
The Schwarzschild Solution

Einstein's field equations are nonlinear and are therefore very complicated, and it is difficult to get accurate solutions of them. There is, however, one special case which can be solved without too much trouble, namely, the static spherically symmetric field produced by a spherically mass M at rest. Schwarzschild first found this in 1916, and this solution has played a major role in the early development of general relativity and is even today regarded as a solution of fundamental importance. In the Newtonian gravitation the solution of this problem is described by a gravitational potential $\Phi = GM/r$, where M is the gravitational mass of the source distribution and r is the radius coordinate from the center of the mass distribution. The Schwarzschild solution describes the general relativistic analog of the Newtonian solution.

4.1 The Schwarzschild Metric

The static condition means that, with a static coordinate system, the g_{ij} are independent of the time x^0 or t and also $g_{oi} = 0$. The spatial coordinates may be taken to be spherical polar coordinates $x^1 = r, x^2 = \theta, x^3 = \varphi$. The most general form for ds^2 compatible with spherical symmetry is

$$ds^2 = A \, dt^2 - B \, dr^2 - C \, r^2 \left(d\theta^2 + \sin^2 \theta \, d\varphi^2 \right), \tag{4.1}$$

where A, B, and C are general functions of r only. We may replace r by any function of r without disturbing the spherical symmetry. We now use this freedom to simplify things as much as possible, and the most convenient arrangement is to have $C = 1$. The expression for ds^2 may then be written in the form

$$ds^2 = e^{2\nu} \, dt^2 - e^{2\lambda} dr^2 - r^2 \left(d\theta^2 + \sin^2\theta \, d\varphi^2 \right). \tag{4.2}$$

The functions ν and λ depend on r only. The Schwarzschild radial coordinate has a special meaning: the area of the surface of the sphere, $r = $ constant, is given by

$4\pi^2 r^2$. In the future we will refer to r as a Schwarzschild radial coordinate whenever we want to imply the above result.

4.2 The Schwarzschild Solution of the Vacuum Field Equations

We now work out the Einstein field equations in detail. We can immediately read off the values of the $g_{\mu\nu}$ from (4.2):

$$g_{00} = e^{2\nu}, \quad g_{11} = -e^{2\lambda}, \quad g_{22} = -r^2, \quad g_{33} = -r^2\sin^2\theta, \qquad (4.3)$$

$$g_{\mu\nu} = 0, \quad \text{for} \quad \mu \neq \nu \qquad (4.4)$$

and we find

$$g^{00} = e^{-2\nu}, \quad g^{11} = -e^{-2\lambda}, \quad g^{22} = -r^{-2}, \quad g^{33} = -r^{-2}\sin^{-2}\theta, \qquad (4.5)$$

$$g^{\mu\nu} = 0, \quad \text{for} \quad \mu \neq \nu. \qquad (4.6)$$

With these values it is easy to calculate the $\Gamma^\alpha{}_{\beta\gamma}$ from the following formula:

$$\Gamma^\alpha{}_{\beta\gamma} = \frac{1}{2}g^{\alpha\delta}\left(\frac{\partial g_{\delta\beta}}{\partial x^\gamma} + \frac{\partial g_{\delta\gamma}}{\partial x^\beta} - \frac{\partial g_{\beta\gamma}}{\partial x^\delta}\right).$$

The calculation leads to the following expressions, with primes denoting differentiations with respect to r,

$$\Gamma^1{}_{00} = \nu' e^{2\nu-2\lambda}, \quad \Gamma^1{}_{11} = \lambda', \quad \Gamma^1{}_{22} = -re^{-2\lambda}, \quad \Gamma^1{}_{33} = -r\sin^2\theta e^{-2\lambda} \qquad (4.7)$$

and

$$\Gamma^0{}_{10} = \nu', \quad \Gamma^2{}_{12} = \Gamma^3{}_{13} = r^{-1}, \quad \Gamma^3{}_{23} = \cot\theta, \quad \Gamma^2{}_{33} = -\sin\theta\cos. \qquad (4.8)$$

All other components (except for those that differ from the ones we have written by a transposition of the indices β and γ) are zero. Substituting these expressions into the Ricci tensor, (2.62)

$$R_{\mu\nu} = \frac{\partial \Gamma^\alpha{}_{\mu\alpha}}{\partial x^\nu} - \frac{\partial \Gamma^\alpha{}_{\mu\nu}}{\partial x^\alpha} - \Gamma^\alpha{}_{\mu\nu}\Gamma^\beta{}_{\alpha\beta} + \Gamma^\alpha{}_{\mu\beta}\Gamma^\beta{}_{\nu\alpha}$$

we obtain

$$R_{00} = \left(-\nu'' + \lambda'\nu' - \nu'^2 - \frac{2\nu'}{r}\right)e^{(2\nu-2\lambda)}, \qquad (4.9)$$

$$R_{11} = \nu'' - \lambda'\nu' + \nu'^2 - \frac{2\lambda'}{r} \qquad (4.10)$$

$$R_{22} = (1 + r\nu' - r\lambda')e^{-2\lambda} - 1, \qquad (4.11)$$

$$R_{33} = R_{22}\sin^2\theta, \qquad (4.12)$$

with the other components of $R_{\mu\nu}$ vanishing.

Einstein's law of gravity in empty space requires these expressions to vanish. The vanishing of (4.9) and (4.10) leads to

$$\lambda' + v' = 0.$$

For large values of r the space must approximate to being flat, so that λ and v both tend to zero as $r \to \infty$. It follows that

$$\lambda + v = 0. \tag{4.13}$$

The vanishing of (4.11) gives

$$\left(1 + 2rv'\right) e^{2v} = 1$$

or

$$\left(re^{2v}\right)' = 1.$$

Thus,

$$re^{2v} = r - 2m \tag{4.14}$$

where m is an integration constant. Now we have

$$g_{00} = 1 - 2m/r. \tag{4.15}$$

The constant m can be expressed in terms of the mass of the body by requiring that Newton's law should hold at large distances where the field is weak. In other words, we should have $g_{00} = 1 + 2\phi/c^2$, where the potential has its Newtonian value $\phi = -GM/r$. From this it is clear that $m = GM/c^2$. This quantity has the dimension of length; it is called the gravitational radius r_g of the body: $r_g = 2GM/c^2$. So,

$$e^{2v} = g_{00} = 1 - 2GM/rc^2 = 1 - r_g/r \tag{4.16}$$

and

$$e^{2\lambda} = e^{-2v} = g_{11} = \left(1 - r_g/r\right)^{-1}. \tag{4.17}$$

Thus, we finally obtain the space-time metric in the form

$$ds^2 = \left(1 - r_g/r\right) c^2 dt^2 - \left(1 - r_g/r\right)^{-1} dr^2 - r^2 \left(d\theta^2 + \sin^2\theta \, d\varphi^2\right) \tag{4.18}$$

where $r = 0$ is the center of the body. The coordinates (t, r, θ, φ) are referred to as the *static curvature coordinates*. Schwarzschild found this solution of the Einstein equations in 1916. It tells us the important result: "the empty space-time outside a spherically symmetric distribution of matter is describable by a static metric." This result is known as *Birkhoff's theorem*.

The Schwarzschild solution is also valid for centrally symmetric distribution of matter that is moving, so long as the motion has the required symmetry, for example a centrally symmetric pulsation. We note that the metric (4.18) depends only on the

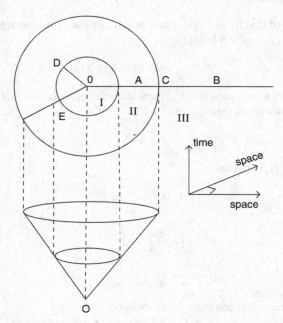

Fig. 4.1 Applicability of the space-time metric (Eq. 4.18) to different regions around a spherically symmetric distribution of matter.

total mass of the gravitating body, just as in the analogous problem in Newtonian theory.

The gravitational radius (also known as the Schwarzschild radius) r_g is a very important constant of integration. In a normal situation the physical radius r_b of the central body is far larger than r_g and the field (4.18) is valid for $r > r_g$ only. For example, for our sun $r_g = 2.9$ km; and for Earth, it is 0.88 cm. But it is conceivable in astronomy that the body could be in such a high state of compression that $r_g > r_b$. This situation is the Schwarzschild black hole, as shown in Figure 4.1, where OC = r_g. The central body occupies region I of radius OD = $r_b < r_g$, where the metric (4.18) does not apply. In region II ($r_g \geq r > r_b$) there is no material and therefore the metric (4.18) might be supposed to apply. In region III, where $r > r_g$ the metric (4.18) applies without difficulty. The boundary between regions II and III ($r = r_g$) is characterized by a zero value of the coefficient of dt^2, and an infinite value of that dr^2, which constitutes the singularity. We will see in Chapter 6 that the singularity at $r = r_g$, is more a consequence of the choice of the coordinate frame than of the geometry itself. Nevertheless, interesting things do happen at the boundary $r = r_g$. For example, matter and energy may fall into the region $r < r_g$, but nothing (including light signal) can come out of it. We will discuss this situation further in Chapter 6.

The singularity at $r = 0$ is an intrinsic one that cannot be eliminated by any transformation of coordinates. The space-time curvature itself is singular there:

$$R^{\mu\nu\alpha\chi} R_{\mu\nu\alpha\beta} = \frac{12}{r^6} \left(\frac{2GM}{c^2}\right)^2.$$

The spatial metric is determined by the expression for the element of spatial distance:

$$-dL^2 = \left(1 - \frac{r_g}{r}\right)^{-1} dr^2 + r^2 d\theta^2 + r^2 \sin^2\theta \, d\varphi^2. \tag{4.19}$$

This allows us to refer to events with the same r, θ, φ-coordinates, but different time coordinates, as occurring at the same point in space. We would like to emphasize that this splitting of space-time into space and time is not a feature of space-time in general; it is possible only in any static space-time.

When r_g is much smaller than r so that r_g/r is negligible, then the line element (4.19) becomes that of flat space in spherical polar coordinates, and the coordinate r is the distance from the origin. In a curved space, r no longer measures radial distance, and the distance between two points r_1 and r_2 along the same radius is given by the integral

$$\int_{r_1}^{r_2} \frac{dr}{\sqrt{1 - r_g/r}} > r_2 - r_1. \tag{4.20}$$

But in the metric (4.19), the circumference of a circle with its center at the center of the field is $2\pi r$. We can see this by considering a sphere of a given radius, then $dr = 0$ in the line element (4.19) and we now have

$$dL^2 = r^2 \left(d\theta^2 + \sin^2\theta d\varphi^2\right), \tag{4.21}$$

which shows that the sphere has the two-dimensional geometry of a sphere of radius r.

We now turn our attention to the measurement of time in the curved space-time. Equation (1.17) gives the connection between the proper time $d\tau$ and the dx^0, the coordinate differential

$$d\tau = \frac{1}{c}\sqrt{g_{00}} \, dx^0 = \sqrt{g_{00}} \, dt. \tag{1.17}$$

Now, $g_{00} = 1 - r_g/r$, and (1.17) becomes

$$d\tau = \left(1 - r_g/r\right)^{1/2} dt \tag{4.22}$$

where $d\tau$ is the proper time of observers at rest at Point r. We see that

$$d\tau \leq dt. \tag{4.23}$$

The equality sign holds only at infinity, where t coincides with the proper time. At finite distances from the masses there is a "slowing down" of the time compared with the time at infinity.

Light of a certain frequency is emitted and is being observed at a large value of r – it is redshifted. To see this, we first compare the proper time intervals evaluated at two distinct points in space but both corresponding to the same intervals of coordinate time dt. The ratio of the proper time intervals, according to (1.17), is

$$\frac{d\tau_1}{d\tau_2} = \sqrt{\frac{g_{00}(x_1)}{g_{00}(x_2)}}. \tag{4.24}$$

The frequency of the same radiation measured by an observer at rest at the two points will be different. The ration of the two frequencies follows from (4.24)

$$\nu_2\sqrt{g_{00}(x_2)} = \nu_1\sqrt{g_{00}(x_1)}. \tag{4.25}$$

Thus, light is redshifted as it propagates away from the gravitating mass. If light is emitted at Point r with frequency ν, then when received "at infinity" it will be redshifted to a frequency ν_∞ with

$$\nu_\infty = \nu\sqrt{1 - r_g/r}. \tag{4.26}$$

Obviously, as $r \to \infty$, $d\tau \to dt$, and the line element dS^2 becomes

$$ds^2 = ds_0{}^2 - \left(2GM/c^2 r\right)\left(dr^2 + c^2 dt^2\right). \tag{4.27}$$

The second term represents a small correction to the Galilean metric $ds_0{}^2$. At large distances from the masses producing it, every field appears centrally symmetric. Therefore, (4.27) determines the metric at large distances from any system of bodies.

4.3 Schwarzschild Geodesics

The paths of particles with mass moving in the vicinity of a spherical massive object are given by the time-like geodesics of space-time, Eq. (2.44):

$$\frac{d^2 x^\nu}{ds^2} + \Gamma^\nu_{\lambda\sigma}\frac{dx^\lambda}{ds}\frac{dx^\sigma}{ds} = 0. \tag{2.44}$$

We now solve these equations in the Schwarzschild metric. We first construct the θ – component of the geodesic equation, $\nu = 2$. From (2.44) and (4.7) and (4.8) we have

$$\frac{d^2\theta}{ds^2} = -\Gamma^2_{\lambda\sigma}\frac{dx^\lambda}{ds}\frac{dx^\sigma}{ds}$$

$$= -2r^{-1}\frac{dr}{ds}\frac{d\theta}{ds} + \sin\theta\cos\theta\left(\frac{d\varphi}{ds}\right)^2. \tag{4.28}$$

The right-hand side vanishes when $\theta = \pi/2$ and $d\theta/ds = 0$. Thus, motion occurs in a plane passing through the origin, exactly as in Newtonian central force motion. We can use the spherical symmetry to perform a suitable orientation of the coordinate axes to ensure that $\theta = \pi/2$ initially; then the motion will remain at the equatorial plane ($\theta = \pi/2$). With this simplification we can easily calculate the φ–and t— components of the geodesic equation:

$$\frac{d^2\varphi}{ds^2} = -\Gamma^3_{\lambda\sigma}\frac{dx^\lambda}{ds}\frac{dx^\sigma}{ds} = -2r^{-1}\frac{d\varphi}{ds}\frac{dr}{ds} \tag{4.29}$$

$$\frac{d^2t}{ds^2} = -\Gamma^0_{\lambda\sigma}\frac{dx^\lambda}{ds}\frac{dx^\sigma}{ds} = -v'\frac{dt}{ds}. \tag{4.30}$$

We rewrite these two equations in forms that are ready for integration:

$$\frac{d}{ds}\left(\ln\frac{d\varphi}{ds}\right) = -2\frac{d}{ds}\ln r \tag{4.31}$$

$$\frac{d}{ds}\left(\ln\frac{dt}{ds}\right) = -\frac{dv}{ds}. \tag{4.32}$$

Integration gives

$$\frac{d\varphi}{ds} = \frac{h}{r^2}, \tag{4.33}$$

and

$$\frac{dt}{ds} = E\left(1 - \frac{r_g}{r}\right)^{-1}, \tag{4.34}$$

where h and E are constants.

Equations (4.33) and (4.34) arise from the symmetries in φ and t respectively. There is an integral of the geodesic equation for each symmetry of the metric. This can be shown easily. First put in the expression (2.42) for the Christoffel symbol in (2.44) and then multiply the equation by $g_{\mu\nu}$ to obtain

$$g_{\mu\nu}\frac{dU^\nu}{ds} + \frac{1}{2}g_{\mu\nu}g^{\nu a}\left(\frac{\partial g_{\sigma a}}{\partial x^\lambda} + \frac{\partial g_{\lambda a}}{\partial x^\sigma} - \frac{\partial g_{\lambda\sigma}}{\partial x^a}\right)U^\lambda U^\sigma = 0,$$

where $U^\nu = \frac{dx^\nu}{ds}$, etc.

Using the relation $g_{\mu\nu}g^{\nu\lambda} = \delta^\lambda_\mu$, the last expression becomes

$$g_{\mu\nu}\frac{dU^\nu}{ds} + \left(\frac{\partial g_{\mu\nu}}{\partial x^\lambda} - \frac{1}{2}\frac{\partial g_{\nu\lambda}}{\partial x^\mu}\right)U^\lambda U^\nu = 0.$$

Now,

$$\frac{dU_\mu}{ds} = \frac{d}{ds}\left(g_{\mu\nu}U^\nu\right) = g_{\mu\nu}\frac{dU^\nu}{ds} + \frac{\partial g_{\mu\nu}}{\partial x^\lambda}U^\lambda U^\nu.$$

Combining this with the last equation we obtain

$$\frac{dU_\mu}{ds} - \frac{1}{2}\frac{\partial g_{\nu\lambda}}{\partial x^\mu}U^\nu U^\lambda = 0. \tag{4.35}$$

Thus, if $\partial g_{\nu\lambda}/\partial x^\mu = 0$ for all ν and λ, U_μ is a constant of the motion, or a first integral. In general, there is an integral of the geodesic equation for each symmetry of the metric.

We next use the fact that the line element (4.18) itself provides a first integral:

$$1 = (1-r_g/r)c^2(dt/ds)^2 - (1-r_g/r)^{-1}(dr/ds)^2 - r^2(d\theta/ds)^2 - r^2\sin^2\theta(d\varphi/ds)^2.$$

Now, $d\theta/ds = 0$, and with (4.33) and (4.34) we get from the last equation:

$$\left(\frac{dr}{ds}\right)^2 = E^2 - \left(1 + \frac{h^2}{r^2}\right)\left(1 - \frac{r_g}{r}\right). \tag{4.36}$$

We are now able to find $r(s)$ from this equation, and then (4.33) and (4.34) could be integrated to give the complete solution to the problem.

Similarly, we could look at the orbits of photons and other zero mass particles in Schwarzschild's geometry. The path of a light ray is a null geodesic; it is fixed by (2.44), but referring to some parameter, say λ, along the path.

4.4 Quasiuniform Gravitational Field

The gravitational field described by the following metric

$$ds^2 = -\left(1 + 2gx/c^2\right)c^2dt^2 - \left(1 + 2gx/c^2\right)^{-1}dx^2 - dy^2 - dz^2 \tag{4.37}$$

has been called "quasiuniform," and it may be thought of as representing flat space-time in a suitably accelerated frame of reference.

The metric (4.37) may be derived from the static Schwarzschild metric (4.18) written in the following form:

$$ds^2 = (1 - 2m/r)c^2dt^2 - (1 - 2m/r)^{-1}dr^2 - r^2\left(d\theta^2 + \sin^2\theta d\varphi^2\right) \tag{4.38}$$

where $m = GM/c^2$. Then let us consider Point O fixed in the coordinate system used in (4.38) at $r = r_0$, and a small region of space around O in which the radial coordinate of any point is $r = r_0 + \varepsilon$, where $|\varepsilon| \ll r_0$. Then we can expand

$$1 - 2m/r = 1 - 2m/r_0 + 2m\varepsilon/r_0^2 + \ldots \approx a^2\left[1 = 2m\varepsilon/(a^2r_0^2)\right],$$

where

$$a^2 = 1 - 2m/r_0$$

Setting

$$\varepsilon/a = x, \text{ whence } dr = d\varepsilon = a\,dx$$

$$at = t', mc^2/\left(ar_0^2\right) = g, \text{ and } r^2 d\theta^2 + r^2 \sin^2 \theta d\varphi^2 = dy^2 + dz^2$$

we see that, to our approximation,

$$1 - 2m/r \approx a^2 \left[1 + 2gx/c^2\right],$$

and the matrix in (4.38) transforms into one having the form of (4.37), with t' in place of t.

We can see that (4.37) has arisen as a local representation of (4.38), replacing the radial Schwarzschild field by a parallel field and altering the time coordinate.

With regard to the parameter g, it is not exactly the Newtonian expression GM/r_0^2. Now $M = mc^2/G$, and the Newtonian expression becomes mc^2/r_0^2. We see the Newtonian g is different from our expression by the factor $1/a$.

4.5 Problems

4.1. Prove the results of (4.7) and (4.8).

4.2. Prove the results of (4.9) to (4.12).

4.3. Calculate the gravitational radius (or Schwarzschild radius) for (a) a galaxy ($M = 10^{11} M_\odot$), and (b) a proton.

4.4. Show that the transformation below puts the Schwarzschild metric (26) into the isotropic form:

$$r = r' \left(1 + \frac{GM}{2c^2 r'}\right)^2$$

$$ds^2 = \left(1 + \frac{GM}{2c^2 r'}\right) \left[dr'^2 + r'^2 \left(d\theta^2 + \sin^2 \theta \, d\phi^2\right)\right]$$

$$- \left(\frac{1 - (GM/2c^2 r')}{1 + (GM/2c^2 r')}\right)^2 c^2 dt^2,$$

4.5. Resolve the clock paradox of special relativity by using the above isotropic metric.

4.6. Find the equations determining the static gravitational field in a vacuum around an axially symmetric body at rest.

References

Dirac PAM (1975) *General Theory of Relativity.* (John Wiley & Sons, New York)
Landau LD, Lifshitz EM (1975) *The Classical Theory of Field.* (Pergamon Press, Oxford UK)

Chapter 5
Experimental Tests of Einstein's Theory

The Schwarzschild metric enables the behavior of test bodies, light rays, and clocks to be predicted, so that Einstein's theory of gravitation can be tested. Some of these tests will be discussed in this chapter. We first discuss the two classical tests: the perihelion shift of Mercury and the deflection of light by gravity, then the time delay of light in a gravitational field (the Shapiro experiment). Finally we will comment on Hulse and Taylor's measurement of gravitational wave decay of the orbit of a binary pulsar system. Their work indirectly confirms the existence of gravitational radiation.

5.1 Precession of the Perihelion of Mercury

According to classical celestial mechanics that is based on Newton's law of gravity, a planet describes a closed elliptical orbit with the sun at one of the two foci. This is a result peculiar to the Newtonian inverse-square law; any small deviation from it will cause a planetary orbit not to be closed and its perihelion will shift. Because of Mercury's high velocity and eccentric orbit, the perihelion position can be accurately determined by observation; the difference between the predicted perihelion shift due to perturbation by other planets and the observed perihelion shift is 43 seconds of arc per century. It is a very small difference, but it is about a hundred times the probable observational error and thus represents a true discrepancy from the very precise predictions of celestial mechanics. This discrepancy has bothered astronomers since the middle of the 19th century. The relativistic Newtonian mechanics gives only 1/6 of the observed centennial precession. We consider this problem in the framework of Einstein's general theory of relativity, i.e., consider the motion of Mercury in the space-time continuum characterized by the Schwarzschild line element.

We assume that the mass of Mercury is negligible in comparison with the mass of the sun ($M_\odot = 2 \times 10^{30}$ kg, $M_{Mercury} = 3.3 \times 10^{23}$ kg), so that its presence does not modify the metric to any appreciable extent. The motion of Mercury is a time-like geodesic in the Schwarzschild space-time surrounding the sun. In classical mechanics the orbit of a body in a central force field lies in a plane. This is true in the present theory. By an appropriate orientation of the axes we can make $\theta = \pi/2$ and $\dot{\theta} = 0$ at some initial s, and it will stay that way for all s. Then, from (4-40a) and (4-41a), we have

$$r^2 \dot{\varphi} = h = \text{constant} \tag{5.1}$$

$$\left(1 - \frac{2GM}{c^2 r}\right) \dot{t} = E = \text{constant} \tag{5.2}$$

where we use dot for d/ds. We now divide the line element (4.18) by ds^2 to obtain a third differential equation:

$$1 = \left(1 - \frac{2GM}{c^2 r}\right) c^2 \dot{t}^2 - \left(1 - \frac{2GM}{c^2 r}\right)^{-1} \dot{r}^2 - r^2 \left(\dot{\theta}^2 + \sin^2 \theta \dot{\varphi}^2\right). \tag{5.3}$$

Substituting (5.1), (5.2), and $\theta = \pi/2$ into (5.3) we get the differential equation for $r(s)$:

$$1 = \left(1 - \frac{2GM}{c^2 r}\right)^{-1} c^2 E^2 - \left(1 - \frac{2GM}{c^2 r}\right)^{-1} \dot{r}^2 - \frac{h^2}{r^2}. \tag{5.4}$$

We can simplify matters by considering r as a function of φ instead of s and we use prime for $d/d\varphi$. Then,

$$r' = \frac{dr}{d\varphi} = \frac{\dot{r}}{\dot{\varphi}}. \tag{5.5}$$

From (5.1) and (5.5) we obtain

$$\dot{r} = \dot{\varphi} r' = \frac{h}{r^2} r'. \tag{5.6}$$

Substituting (5.6) into (5.4) we obtain the differential equation for $r(\varphi)$:

$$\left(1 - \frac{2GM}{c^2 r}\right) = c^2 E^2 - \frac{h^2}{r^4} r'^2 - \frac{h^2}{r^2} \left(1 - \frac{2GM}{c^2 r}\right). \tag{5.7}$$

Now let

$$r = 1/u \tag{5.8}$$

then

$$r' = -u'/u^2$$

and (5.7) converts into a differential equation for $u(\varphi)$:

$$\left(1 - \frac{2GM}{c^2} u\right) = c^2 E^2 - h^2 u'^2 - h^2 u^2 \left(1 - \frac{2GM}{c^2} u\right)$$

or

$$u'^2 = \left(\frac{c^2 E^2 - 1}{h^2}\right) + \frac{2GM}{c^2 h^2} u - u^2 + \frac{2GM}{c^2} u^3, \tag{5.9}$$

which can be integrated in principle, giving the angle φ as an integral of u. But it cannot be integrated in terms of elementary functions. The enterprising student may wish to attempt the integration by use of elliptical functions. Instead, we shall convert the first-order equation (5.9) to a second-order equation by differentiation with respect to φ. This will make the problem more transparent and establish a closer connection with the classical Kepler problems that involve a second-order differential equation. Differential (5.9) with respect with φ, we obtain

$$2u'u'' = \frac{2GM}{c^2h^2}u' - 2uu' + \frac{6GM}{c^2}u^2u' = 0. \tag{5.10}$$

One possible solution is obtained by setting the common factor u' equal to zero:

$$u' = 0, \quad u = \text{constant} \quad r = \text{constant}.$$

Thus circular motion occurs in relativity theory just as in classical theory. The other more interesting solution will result from canceling the common factor u' from (5.10):

$$u'' + u = \frac{GM}{c^2h^2} + \frac{3GM}{c^2}u^2. \tag{5.11}$$

This equation is very similar in structure to the orbit equation of the classical Kepler problem:

$$u'' + u = GM/H^2$$

where prime stands for d/dt and H is twice the constant area velocity

$$H = r^2\frac{d\varphi}{dt} = \text{constant}.$$

(5.11) differs from the classical Kepler problem by the presence of the addition of the quadratic term $(3GM/c^2)u^2$. The constant term GM/c^2h^2 is slightly different:

$$\frac{GM}{c^2h^2} = \frac{GM}{c^2r^4(d\varphi/ds)^2} = \frac{GM}{c^2r^4(d\varphi/dt)^2(dt/ds)^2}.$$

For slowly moving bodies in weak gravitational fields, (dt/ds) is approximately equal to $1/c$; substituting this in the last equation we obtain

$$\frac{GM}{c^2h^2} = \frac{GM}{c^2r^4(d\varphi/ds)^2} = \frac{GM}{c^2r^4(d\varphi/dt)^2(dt/ds)^2} \cong \frac{GM}{r^4(d\varphi/dt)^2} = \frac{GM}{H^2}.$$

We now show that the quadratic term is small relative to the leading constant term:

$$\frac{3GMu^2/c^2}{GM/c^2h^2} = 3u^2h^2 = 3r^2\dot{\varphi}^2 \cong 3[r(d\varphi/dt)]^2 \cdot 1/c^2.$$

The quantity $r(d\varphi/dt)$ is the lateral velocity of the planet v_r (the velocity perpendicular to r), so the above ratio can be written as $3(v_r)^2/c^2$, which is very small.

Thus we may consider the quadratic term in (5.11) as a small perturbation term; and this allow us to take a perturbation approach. Define

$$A = \frac{GM}{c^2 h^2} \cong \frac{GM}{c^2 H^2} \tag{5.12}$$

and

$$\varepsilon = \frac{3GM}{c^2} A \cong \frac{3G^2 M^2}{c^4 H^2} \tag{5.13}$$

Equation (5.11) now takes the form

$$u'' + u = A + \frac{\varepsilon}{A} u^2. \tag{5.14}$$

To solve this equation we assume a solution of the form

$$u(\varphi) = u_0(\varphi) + \varepsilon v(\varphi) + O\left(\varepsilon^2\right) \tag{5.15}$$

and attempt to find $u_0(\varphi)$ and $v(\varphi)$.

Substituting (5.15) in the differential equation (5.14), we obtain

$$u_0'' + \varepsilon v'' + u_0 + \varepsilon v = A + \varepsilon u_0^2 / A + O\left(\varepsilon^2\right) \tag{5.16}$$

Equating the zeroth-order term in ε, we obtain

$$u_0'' + u_0 = A. \tag{5.17}$$

Its solution is

$$u_0 = A + B \cos(\varphi + \delta)$$

where B and δ are arbitrary integration constants. By an appropriate orientation of the axes we may make δ equal to zero. Then we have

$$u_0 = A + B \cos \varphi \tag{5.18}$$

Now, equating the first-order ε term in Eq. (5.16), we obtain

$$v'' + v = u_0^2 / A = \left(A + \frac{B^2}{2A}\right) + 2B \cos \varphi + \frac{B^2}{2A} \cos 2\varphi. \tag{5.19}$$

Note that we need only a nonhomogeneous solution to this equation since the zeroth-order solution already contains a term $B \cos \varphi$, which is the general solution to the homogeneous equation. Since (5.19) is linear in v, we may write v as the sum $v = v_a + v_b + v_c$, where

$$v_a'' + v_a = A + \frac{B^2}{2A} \quad v_b'' + v_b = 2B \cos \varphi \quad v_c'' + v_c = \frac{B^2}{2A} \cos 2\varphi. \tag{5.20}$$

The nonhomogeneous solutions to (5.20) are easily checked to be

$$v_a = A + \frac{B^2}{2A} \quad v_b = B\varphi \sin \varphi \quad v_c = -\frac{B^2}{6A} \cos 2\varphi.$$

Then,

$$v = v_a + v_b + v_c = A + \frac{B^2}{2A} + B\varphi \sin \varphi - \frac{B^2}{6A} \cos 2\varphi, \qquad (5.21)$$

which is a nonhomogeneous solution to (5.19). Combining this with (5.18) we have the solution for the orbit to the first order in ε:

$$u = u_0 + \varepsilon v = \left(A + \varepsilon A + \frac{\varepsilon B^2}{2A} \right) + \left(B \cos \varphi - \frac{\varepsilon B^2}{6A} \cos 2\varphi \right) + \varepsilon B\varphi \sin \varphi. \quad (5.22)$$

Note that only the last is nonperiodic; any irregularities occurring in the perihelion position are due to this term. We also note that, to first order in ε:

$$\cos(\varphi - \varepsilon\varphi) = \cos \varphi \cos \varepsilon\varphi + \sin \varphi \sin \varepsilon\varphi = \cos \varphi + \varepsilon\varphi \sin \varphi.$$

The solution (5.22) now may be written as

$$u = A + B \cos(\varphi - \varepsilon\varphi) + \varepsilon \left(A + \frac{B^2}{2A} - \frac{B^2}{6A} \cos 2\varphi \right). \qquad (5.23)$$

The last term introduces a small periodic variation in the radial distance of the planet, which is difficult to detect and will not influence the perihelion motion. But the $\varepsilon\varphi$ that appears in the cosine argument will introduce a nonperiodicity, and since φ can become large, the effect is not negligible. Accordingly, we now write (5.23) in the form

$$u = A + B \cos(\varphi - \varepsilon\varphi) + (periodic\ terms\ of\ order\ \varepsilon). \qquad (5.24)$$

The perihelion of a planet occurs when r is a minimum or when $u(= 1/r)$ is a maximum. From (5.24) we see that u is a maximum when

$$\varphi(1 - \varepsilon) = 2n\pi \qquad (5.25)$$

or approximately

$$\varphi = 2n\pi(1 + \varepsilon). \qquad (5.26)$$

Therefore, successive perihelia will occur at intervals of

$$\Delta\varphi = 2\pi(1 + \varepsilon) \qquad (5.27)$$

and the perihelion shift per revolution is given by $\delta\varphi$ (see Fig. 5.1)

$$\delta\varphi = 2\pi\varepsilon = 2\pi \frac{3G^2 M^2}{c^4 H^2}. \qquad (5.28)$$

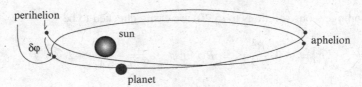

Fig. 5.1 The shift of the perihelion.

It is more convenient to express H in terms of the major axis and eccentricity e of the ellipse. The length of the major axis, denoted by $2a$, is given by

$$2a = \left(\frac{1}{u_0}\right)_{\varphi=0} + \left(\frac{1}{u_0}\right)_{\varphi=\pi}. \tag{5.29}$$

Using this relation, we can write H^2 in terms of the eccentricity e and semimajor axis a:

$$H^2 = \frac{aGM}{c^2}\left(1 - e^2\right). \tag{5.30}$$

The perihelion shift per revolution may now be written as

$$\delta\varphi = \frac{6\pi\,GM}{c^2 a\left(1 - e^2\right)} \text{ radians.} \tag{5.31}$$

The dependence of the perihelion advance on the eccentricity e and the semimajor axis a is evident. In Table 5.1, we list observational data for Mercury, Venus, Earth, and Icarus (asteroid); their semimajor axes are small enough and M large enough for $\delta\varphi$ to be measured. Shown is the perihelion advance per century. For Mercury, the period of its orbit is 0.241 Earth days, so in Earth century, Mercury completes about 415 orbits. The total advance, expressed in seconds of arc rather than radians, is $43.03''$ per century.

Table 5.1 Observational Data for Selected Planets

Planet	a ($10^6\,km$)	e	φ (seconds of arc per century)	
			Observed	Theoretical
Mercury	57.91	0.2056	43.11 ± 0.45	43.03
Venus	108.21	0.0068	8.4 ± 4.8	8.6
Earth	149.60	0.0167	5.0 ± 1.2	3.8
Icarus	161.0	0.827	9.8 ± 0.8	10.3

The precision of these observations is quite remarkable when we consider that for the planet Mercury, the measured perihelion shift per century amounts to

$$(\text{Mercury}) = 5600.73 \pm 0.41''.$$

Of this value, the shift caused by known nonrelativistic disturbance effects is

$$(\text{Disturbance}) = 5557.62 \pm 0.20''.$$

` 43.11 $\pm$ 0.45″ is attributed to relativistic effect. Astronomers knew the advance of Mercury's perihelion as early as 1860. Its final explanation through the general theory of relativity in 1915 was one of the greatest triumphs of Einstein's theory.

5.2 Deflection of Light Rays in a Gravitational Field

In this section we shall treat a second interesting case of motion in the sun's gravitational field, the trajectory of a light ray. As with Mercury's perihelion shift, the predictions can be subjected to observational testing within the solar system.

As with the case of a massive test particle, the propagation of light rays in a gravitational field follows a geodesic line in a Riemann space. In special relativity the path of a light ray that lies on the light cone is characterized in space-time by its null line element, $ds^2 = 0$. We assume that the same is true in general relativity. Thus, in short, the light-ray trajectories are null-geodesic lines.

When discussing null geodesics we must observe that the curve parameter s that we have been using until now is no longer admissible since $s = 0$ holds on null geodesics; it is replaced by some parameter, say λ, along the path. In the case of the Schwarzschild metric, we find the equations of motion for φ and t as before:

$$r^2 \dot{\varphi} = h = \text{constant} \tag{5.1}$$

$$\left(1 - \frac{2GM}{c^2 r}\right) \dot{t} = E = \text{constant}. \tag{5.2}$$

The dots now denote differentiation with respect to λ, and we have selected as before $\theta = \pi/2$. Instead of (5.4), we now have, since $ds^2 = 0$,

$$0 = \left(1 - \frac{2GM}{c^2 r}\right)^{-1} c^2 E^2 - \left(1 - \frac{2GM}{c^2 r}\right)^{-1} \dot{r}^2 - \frac{h^2}{r^2}. \tag{5.32}$$

Thus, proceeding as before with the substitution $u(\varphi) = 1/r(\varphi)$, we obtain

$$0 = c^2 E^2 - h^2 u'^2 - h^2 u^2 \left(1 - \frac{2GM}{c^2} u\right). \tag{5.33}$$

Next, by differentiation (5.33) with respect to φ, we have

$$u' \left(u'' + u - \frac{3GM}{c^2} u^2\right) = 0. \tag{5.34}$$

The equation for a light-ray trajectory is given by

$$u'' + u = \frac{3GM}{c^2} u^2. \tag{5.35}$$

The other special solution is $u' = 0$, or $u = \text{constant}$. This solution would describe light rays circling the attracting center at a fixed distance $r = r_0$. Such singular solutions occurred also in the theory of planetary motion, and in that theory they have physical reality. The situation in the present case is different. Observe that the general (5.11) admits $u = u_0$ as a solution for an appropriate choice of the initial angular momentum. However, $u = u_0$ is a solution of the light-ray equation (5.35) only if $u_0^{-1} = r_0 = 3GM/c^2$. Hence the singular solutions of $u' = 0$ cannot be changed continuously into solutions of the more general equation (35) except at $r_0 = 3GM/c^2$. Thus these solutions are in general unstable.

As with the orbit equation of the preceding section, the term $(3GM/c^2)u^2$ is small relative to the other terms of the equation. To do this, let us consider the ratio of $(3GM/c^2)u^2$ to the term u; that is, consider $(3GM/c^2)u$. Using the definition of the Schwarzschild radius r_g, we may also write this ratio as $(3/2)(r_g/r)$. The Schwarzschild radius of the sun is of the order of a kilometer; thus, for a trajectory outside the sun's surface, the above ratio is evidently very small. This allows us to regard $(3GM/c^2)u^2$ as a small perturbation term in (5.35). Accordingly, let us call

$$\frac{3GM}{c^2} = \varepsilon$$

and (5.35) becomes

$$u'' + u = \varepsilon u^2. \tag{5.36}$$

As in the preceding section, we shall use a standard perturbation approach to treat the above equation; we suppose a solution to (5.15) of the form

$$u = u_0 + \varepsilon v + O\left(\varepsilon^2\right). \tag{5.37}$$

Substituting this in (5.36), we have

$$u_0'' + u_0 + \varepsilon v'' + \varepsilon v = \varepsilon u_0^2 + O\left(\varepsilon^2\right). \tag{5.38}$$

Equating the zeroth-order terms in ε, we have

$$u_0'' + u_0 = 0. \tag{5.39}$$

This has the solution

$$u_0 = A \sin(\varphi - \alpha), \tag{5.40}$$

which, by an appropriate orientation of the axes, may be written without the arbitrary constant α

$$u_0 = A \sin \varphi \tag{5.41}$$

or

$$r \sin \varphi = \frac{1}{A}. \tag{5.42}$$

Note that $r \sin \varphi$ is simply the Cartesian coordinate y, a straight line parallel to x axis. This is what we should expect; in first approximation the light ray is not deflected by the sun's gravitational field. (5.42) also indicates that the distance of closest approach to the origin (the sun) is $1/A$ and denotes it by r_0. Thus we can write the zeroth-order solution as

$$u_0 = \frac{1}{r_0} \sin \varphi. \tag{5.43}$$

Next, equating the first-order ε terms of (5.38), we obtain

$$v'' + v = u_0^2 = \frac{1}{r_0^2} \sin^2 \varphi = \frac{1}{2r_0^2}(1 - \cos 2\varphi). \tag{5.44}$$

We now try a solution with two unknown coefficients

$$v = \alpha + \beta \cos 2\varphi. \tag{5.45}$$

Differentiation gives

$$v'' = -4\beta \cos 2\varphi$$

so that

$$v'' + v = \alpha - 3\beta \cos 2\varphi. \tag{5.46}$$

Comparing this equation term by term with (5.44), we see that

$$\alpha = \frac{1}{2r_0^2} \beta = \frac{1}{6r_0^2} \tag{5.47}$$

and the solution of (5.44) is

$$v = \frac{1}{2r_0^2}\left(1 + \frac{1}{3}\cos 2\varphi\right). \tag{5.48}$$

The full first-order solution to the trajectory (5.36) is

$$u = \frac{1}{r_0}\sin \varphi + \varepsilon \frac{1}{2r_0^2}\left(1 + \frac{1}{3}\cos 2\varphi\right). \tag{5.49}$$

As we have seen above, the trajectory of a light ray as given by (5.39) is essentially a straight line [$u = (1/r_0)\sin \varphi$] with a perturbation of order ε. The effect of this perturbation will alter the trajectory to produce a small overall deflection; that is, light approaches the sun along an asymptotic straight line, is deflected by the gravitational field, and recedes again on another asymptotic straight line. The total deflection can be measured observationally for the case of starlight grazing the sun and finally arriving on Earth. Let us therefore see what total deflection is predicted by (5.49) for such a situation.

The asymptotes of the trajectory will clearly correspond to those values of the angle φ for which r becomes infinite or u becomes zero in (5.49). These asymptotes are nearly parallel to the x axis and correspond to φ being close to zero or π. Thus considering the asymptote near $\varphi = 0$ first and calling δ the small angle between it and the x axis, we approximate $\sin \varphi$ by δ and $\cos 2\varphi$ by 1. Then setting $u = 0$ in (5.49), we obtain

$$0 = \frac{1}{r_0}\delta + \varepsilon \frac{4}{3}\frac{1}{2r_0^2}$$

or

$$\delta = -\frac{2\varepsilon}{3r_0} = -\frac{6GM}{c^2 r_0}. \tag{5.50}$$

The minus sign indicates that the light ray is bent inward by the sun. A similar procedure for the other asymptote, for which φ is taken to be $\pi - \delta$, yields the same value, $\delta = 2GM/c^2 r_0$. Thus the total deflection of light ray, the angle between the asymptotes, is

$$\Delta = \frac{4GM}{c^2 r_0}. \tag{5.51}$$

The deflection is inversely proportional to the distance of closest approach to the origin (the sun); since $M/r_0 \ll 1$ always holds in the solar system the effect is very small. For a light ray that just grazes the sun (so that we take the radius of the sun to be $r_0, r_0 = 6.96 \times 10^{10}$ cm.), (5.51) predicts a deflection of $1.75''$. The early attempts to compare this prediction with observational data utilized photographs taken during solar eclipses. The positions of stellar images near the sun during an eclipse were compared with the positions six months later and with the sun no longer in the field of view. This procedure is inherently difficult since very small displacements of the images have to be measured. As a result the observational results obtained have ranged from $1.47''$ to nearly $2.7''$. With the advent of large radio telescopes and the discovery of the point like sources of intense radio emission called *quasars* the deflection can now be measured using long-baseline interferometric techniques when such a source passes near the sun. Measurements range from 1.57 to $1.82''$, each with an accuracy of about $0.2''$. It is expected that in time this error will be about $0.01''$, resulting in an extremely good agreement with the theory. In very strong gravitational fields, (5.51) is no longer applicable.

If we look at the form of the complete family of curves given by (5.49), which we can evidently interpret as the light from a distant source, then one sees clearly that the light rays converge and produce a caustic line on the axis $\varphi = 0$ (Figure 5.3).

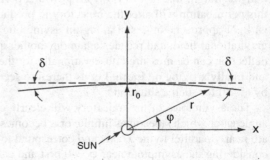

Fig. 5.2 The light ray is bent inward by the sun.

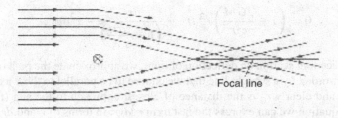

Fig. 5.3 A spherically symmetric gravitational field acts as a gravitational lens.

A spherical symmetric gravitational field thus acts as a gravitational lens. Particularly in the neighborhood of the focal line, interference effects are to be expected.

Gravitational lensing is now regularly observed by astrophysicists, including the lensing of distant quasars by galaxies, and lensing of stars in the galactic nucleus and in the Larger Magellanic Cloud by more nearby stars. About half a dozen quasars seem to have nearby companion quasars with essentially identical spectra but slightly different brightnesses and shapes. The existence of two so nearly identical objects so close together is unlikely. The "companions" are actually images of a single quasar created by a gravitational lens.

5.3 Light Retardation (The Shapiro Experiment)

When passing through a gravitational field, light is retarded, not just deflected. As a result there is a time delay for a light ray traveling close to the sun. I. I. Shapiro showed that such an effect could be measured by sending a radar wave to Venus, where it was reflected back to Earth. The position of Venus has to be in the superior conjunction on the other side of the sun from Earth as shown in Figure 5.4. In this path, the light signal passed close to the sun and experienced a time delay. To calculate this time delay, we use the Schwarzschild metric to calculate the coordinate speed of radar signal traveling in the (r, φ) plane defined by $\theta = \pi/2$.

For radar signal, $ds^2 = 0$, so we have

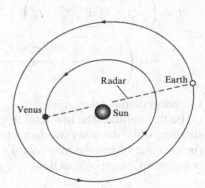

Fig. 5.4 Light retardation by sun's gravitational field acts as a gravitational lens.

$$0 = \left(1 - \frac{2GM}{c^2 r}\right) c^2 dt^2 - \frac{dr^2}{\left(1 - \frac{2GM}{c^2 r}\right)} - r^2 d\varphi^2. \tag{5.52}$$

As in the preceding section on deflection of light, we approximate the path of signal, in the zeroth-order, by a straight line that we choose to be parallel to the x axis. Then $r \sin\varphi = r_0$, and clearly r_0 is the distance of closest approach to the sun (Fig. 5.2). From this equation we can express the last term $r^2 d\varphi^2$ in terms of r and dr:

$$d\varphi = -\frac{\sin\varphi}{\cos\varphi} \frac{dr}{r}, \quad \text{and} \quad d\varphi^2 = \frac{\sin^2\varphi}{\cos^2\varphi} \frac{dr^2}{r^2}$$

$$r^2 d\varphi^2 = \frac{r^2 \sin^2\varphi}{\cos^2\varphi} \frac{dr^2}{r^2} = \frac{r_0^2}{1 - \sin^2\varphi} \frac{dr^2}{r^2} = \frac{r_0^2 dr^2}{r^2 - r_0^2}.$$

Substituting the last equation in (5.52), we obtain

$$c^2 dt^2 = \frac{dr^2}{(1 - 2m/r)^2} + \frac{r_0^2 dr^2}{(1 - 2m/r)\left(r^2 - r_0^2\right)}$$

or

$$c^2 dt^2 = \frac{dr^2 \left(1 - 2m r_0^2/r^3\right)}{(1 - 2m/r)^2 \left(1 - r_0^2/r^2\right)} \tag{5.53}$$

where

$$m = \frac{2GM}{c^2}.$$

Taking the square root and expanding to obtain cdt to first order, we obtain

$$cdt = \frac{dr}{\sqrt{1 - r_0^2/r^2}} \left(1 + \frac{2m}{r} - \frac{m r_0^2}{r^3}\right). \tag{5.54}$$

Integration gives

$$ct = \left(\sqrt{r_p^2 - r_0^2} + \sqrt{r_e^2 - r_0^2}\right) + 2m \ln \frac{\left(\sqrt{r_p^2 - r_0^2} + r_p\right)\left(\sqrt{r_e^2 - r_0^2} + r_e\right)}{r_0^2}$$

$$-m \left(\frac{\sqrt{r_p^2 - r_0^2}}{r_p} + \frac{\sqrt{r_e^2 - r_0^2}}{r_e}\right). \tag{5.55}$$

The integration is taken from $r = r_0$ to r_p, the planet radius, and from $r = r_0$ to r_e (Earth's radius). The first term in (5.55) is the flat-space result for Earth-planet distance, while the other two terms represent an effective increase in the distance. For the solar system we may regard r as a very reasonable radial coordinate and t as an approximate physical time.

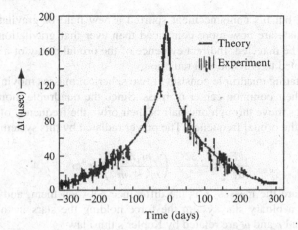

Fig. 5.5 Shapiro's experiments confirm light retardation by sun's gravitational field.

Shapiro and his co-workers have performed several experiments; in every case the predictions of general relativity are confirmed (with an uncertainty of 3%). Figure 5.5 contains the results for the superior conjunctions of Venus in 1970. Several experiments have been done with spacecraft to measure this effect. The experiment to land part of the *Viking* spacecraft on Mars in 1976 produced the most notable measurement. It produced an agreement with theory to within the experimental uncertainty of about 0.1%.

5.4 Test of Gravitational Radiation (Hulse-Taylor's Measurement of the Orbital Decay of the Binary Pulsar PSR-1913+16)

The General Theory of Relativity predicts that gravitational wave radiation will be produced when mass is accelerated, much as electromagnetic radiation is produced when an electric charge accelerates. Gravitational waves carry energy and momentum, travel at the speed of light; and waves interact with all forms of matter. Physicists believe that a particle called the *graviton* mediates the gravitational interaction. This is similar in principle to the photon that mediates the electromagnetic interaction.

It seems unlikely that gravitational waves produced in the laboratory could be detected, but possibly waves from astronomical objects could. These include the collapse of two neutron stars rotating around each other, a neutron star falling into a black hole, and the gravitational collapse of a star to form a black hole. Since the expected wave amplitudes are very small, they were not taken seriously until 1969, when Joseph Weber announced that he had detected gravitational radiation from space. Subsequent investigations by other experimenters have not confirmed

Weber's result, but his announcement spurred a new field of gravitational astrophysics. Scientists are now more convinced than ever that gravitational radiation will eventually be detected. Indirect evidence of the orbital decay of a neutron star has been attributed to gravitational radiation.

A simple rotating quadruple consists of two spherical masses moving in circular orbits around their common center of mass. Since the quadruple moment repeats when the masses move through one-half of their orbit, the frequency of the emitted waves is twice the orbital frequency. The power radiated by this system is

$$\frac{dE}{dt} = -\frac{32G}{5c^5}\left(\frac{m_1 m_2}{m_1 + m_2}\right)^2 r^4 \omega^6 \qquad (5.56)$$

where m_1, m_2 are the masses, r is the difference between them, and ω is orbital frequency. For a binary star system the force holding the stars in their orbits is gravitational, and r and ω are related by Kepler's third law

$$\omega^2 = \frac{G(m_1 + m_2)}{r^3}.$$

Substituting this in (5.56) we obtain

$$\frac{dE}{dt} = -\frac{32G^4}{5c^5 r^5}\left(m_1 m_2\right)^2 (m_1 + m_2). \qquad (5.57)$$

As the binary system loses energy by radiation, the distance between the stars decrease at a rate of

$$\frac{dr}{dt} = -\frac{64G^3}{5c^5 r^3} m_1 m_2 (m_1 + m_2) \qquad (5.58)$$

and the orbital frequency increases at a rate of

$$\frac{d\omega}{dt} = -\frac{3\omega}{2r}\frac{dr}{dt} = \frac{96}{5}\frac{G^{5/2} m_1 m_2 (m_1 + m_2)^{3/2}}{c^5 r^{11/2}}. \qquad (5.59)$$

Thus, as the component stars of a binary system orbit one another, energy escapes in the form of gravity waves, and the two stars slowly spiral toward one another, orbiting more rapidly and emitting even more gravitational radiation. This runaway situation can lead to the decay and eventual merger of close binary systems in a relatively short time (ten or hundreds of millions of years).

Such a slow but steady decay in the orbit of a binary system has in fact been detected. In 1974 Joseph Taylor and his associate Russell Hulse using the Arecibo radio telescope discovered a very unusual binary system, PSR 1913 + 16. Both members are neutron stars, and one is observable from Earth as a pulsar with a pulse period of 59 milliseconds. Measurements of the periodic Doppler shift of the pulsar's radiation prove that its orbit is slowly shrinking. Table 5.2 lists some observed parameters of PSR 1913 + 16.

The orbital period is decreasing by 2.4×10^{-12} second per second. To calculate the theoretical value, we need to know the masses of the two neutron stars. Now, the periastron precession depends only on the sum of $m_1 + m_2$, whereas the time dilation depends on a different combination of m_1 and m_2.

From the measured values of the periastron precession and time dilation Taylor found that

$$m_1 = (1.442 \pm 0.003)M_\odot \quad \text{for pulsar}$$
$$m_2 = (1.386 \pm 0.003)M_\odot \quad \text{for companion}$$

With these values of the masses, the theoretical prediction for the rate of change of the period is 2.38×10^{-12} second per second. We see that the rate at which the orbit is shrinking is exactly what would be predicted by relativity theory if the energy were being carried off by gravitational waves. Even though the waves themselves have not yet been detected, most astronomers regard this binary pulsar as a very strong piece of evidence in favor of general relativity. Taylor and Hulse received the 1993 Nobel Prize in physics for their work.

Table 5.2 Some Observed Parameters of PSR 1913 + 16

Pulsar period (nominal)	$0.059029995271 \pm 0.000000000002$ s
Projected semi-major axis	2.3418 ± 0.0001 light-seconds
Eccentricity	0.617127 ± 0.000003
Orbital period	27906.98163 ± 0.00002 s
Rate of precession of periastron	4.2263 ± 0.0003^0 per year
Amplitude of time-dilation factor	0.0044 ± 0.0001
Rate of change of orbital period	$(-2.40 \pm 0.09) \times 10^{-12}$ s per s

5.5 Problems

5.1. Verify the results of (5.2) to (5.5).

5.2. Calculate the radial coordinate at which light travels in a circular path around a body of mass, using (a) Newtonian mechanics and (b) general relativity. Express the answers as a multiple of the Schwarzschild radius.

5.3. Considering the light-photon as a projectile moving under the influence of Newtonian gravity, calculate the bending of light produced by a massive body. Show that the net bending is half that given by general relativity.

References

Adler R, Bazin M, Schiffer M (1975) *Introduction to General Relativity.* (McGraw-Hill, New York)

Liebes S (1964) Gravitational lenses. Phys Rev B 133: 835–44

Nandor MJ, Helliwell TM (1996) Fermat's principle and multiple imaging by grav-
    itational lenses. Am J Phys 64(1)

Ohanian HC (1987) The black hole as a gravitational lens. Am J Phys 55(5)

Ohanian HC, Ruffini R (1994) *Gravitation and Space-time.* (WW Norton &
    Company, New York)

Shapiro I, Ash M, Campbell D, Dyce R, Ingalls R, Jurgens R, Pettingill, G (1971)
    Fourth test of general relativity: new radar results. Phys Rev Lett, 26: 1132–1135

Stuckey WM (1993) The Schwarzschild black hole as a gravitational mirror. Am J
    Phys 61(5)

Chapter 6
The Physics of Black Holes

In this chapter we mainly study the Schwarzschild black holes, uncharged nonrotating black holes. After 1995 new observational capabilities led us to a new golden age of research on black holes. We learned that the actual behavior of black holes is far more interesting than astrophysicists had imagined. This new information forced them to reexamine many of their tacit assumptions. It is beyond the scope of this book to study all the new exciting developments. Only super massive black holes will be discussed; these may be the central engines of active galaxies.

6.1 The Schwarzschild Black Hole

We noticed in Chapter 4 that in the Schwarzschild solution the metric coefficients g_{11} (i.e., g_{rr}) becomes infinite if r equals the gravitational radius $r_g (= 2GM/c^2)$ and a black hole is formed. Matter may fall into black holes, but nothing (including light signals) can come out of a black hole. We must now prove the truth of these startling assertions. To this aim let us consider a particle falling radially into the central body with the particle having a velocity vector of $v^1 = dx/ds$. Since the particle falls in radially, we can take $v^2 = v^3 = 0$. The motion can be described by the geodesic equation

$$\frac{dv^\mu}{ds} + \Gamma^\mu{}_{\nu\sigma} v^\nu v^\sigma = 0, \tag{6.1}$$

which reduces to, for the case we are considering

$$\frac{dv^0}{ds} = -\Gamma^0{}_{\mu\nu} v^\mu v^\nu = -g^{00}\Gamma_{0,\mu\nu} v^\mu v^\nu = -2g^{00}\Gamma_{0,10} v^o v^1.$$

From

$$\Gamma_{\mu,\nu\sigma} = \frac{1}{2}(g_{\mu\nu,\sigma} + g_{\mu\sigma,\nu} - g_{\nu\sigma,\mu})$$

we find

$$\Gamma_{0,10} = \frac{1}{2}g_{00,1} = \frac{1}{2}\frac{\partial g_{00}}{\partial x^1}.$$

So (6.1) becomes

$$\frac{dv^0}{ds} = -g^{00}\partial\, g_{00,1}v^0\frac{dx^1}{ds} = -g^{00}\frac{dg_{00}}{ds}v^0.$$

Now, $g^{00} = 1/g_{00}$, so we finally get

$$g_{00}\frac{dv^0}{ds} + \frac{dg_{00}}{ds}v^0 = \frac{d}{ds}\left(g_{00}v^0\right) = 0.$$

This integrates to

$$g_{00}v^0 = k, \tag{6.2}$$

with k an integration constant (the value of g_{00} where the particle starts to fall). And from

$$ds^2 = g_{\mu\nu}dx^\mu dx^\nu$$

we have

$$1 = g_{\mu\nu}v^\mu v^\nu = g_{00}\left(v^0\right)^2 + g_{11}\left(v^1\right)^2.$$

Multiplying this equation by g_{00} we obtain

$$g_{00} = \left(g_{00}\right)^2\left(v^0\right)^2 + g_{00}g_{11}\left(v^1\right)^2. \tag{6.3}$$

Now, from Chapter 6, we have

$$g_{00}g_{11} = -e^{2(\lambda+\nu)} = -e^0 = -1.$$

Substituting this and (6.2) into (6.3) we get

$$k^2 - (v^1)^2 = g_{00} = 1 - r_g/r$$

from which we obtain

$$(v^1)^2 = k^2 - 1 + r_g/r. \tag{6.4}$$

For a falling body $v^1 < 0$, and hence

$$v^1 = -\sqrt{k^2 - 1 + r_g/r}. \tag{6.4a}$$

Now, let us consider dt/dr

$$\frac{dt}{dr} = \frac{dx^0/ds}{dx^1/ds} = \frac{v^0}{v^1}$$

and from (6.2) we have

$$v^0 = k/g_{00} = k(1 - r_g/r)$$

so

$$dt/dr = v^0/v^1 = -k\left(1 - r_g/r\right)^{-1}\left(k^2 - 1 + r_g/r\right)^{-1/2}. \tag{6.5}$$

Let us now suppose the particle is close to the critical radius r_g, so $r = r_g + \varepsilon$, with ε small, and let us neglect ε^2. Then

$$dt/dr = -k[1 - (1 + \varepsilon/r_g)^{-1})^{-1}\left[k^2 - 1 + \left(1 + \varepsilon/r_g\right)^{-1}\right]^{-1/2}$$

$$= -k(\varepsilon/r_g)^{-1}\left(k^2 + \varepsilon/r_g\right)^{-1/2}$$

$$= -\left(\varepsilon/r_g\right)^{-1}$$

or

$$dt/dr = -r_g/\varepsilon = -r_g/(r - r_g). \tag{6.6}$$

This integrates to

$$t = -r_g \log(r - r_g) + \text{const.} \tag{6.7}$$

Thus, as $r \to r_g$ and $t \to \infty$, and the particle takes an infinite time to reach the gravitational radius r_g. The surface defined by $r = r_g$ is called the *event horizon*.

The surface area of a Schwarzschild black hole is $4\pi(2GM/c^2)^2$, an expression analogous to the familiar $4\pi r^2$ for the surface of a sphere of radius r. Acceleration of g of a freely falling body near a Schwarzschild black hole is given by

$$g = \frac{GM/r^2}{(1 - 2GM/c^2r)^{1/2}}.$$

When $r \to r_g$, the gravitational force becomes infinite. An infinite quantity is not very convenient to work with; hence, it is convenient to define the "surface gravity" that is just the numerator of the above formula evaluated at the event horizon:

$$\text{Surface gravity} = GM/r_g^2 = c^4/4GM.$$

Both surface area and surface gravity are very useful later when we discuss the thermodynamics of black holes.

We now consider an adventurer traveling with the particle. His time is measured by ds. Now

$$ds/dr = 1/v^1 = -\left(k^2 - 1 + r_g/r\right)^{-1/2} \tag{6.8}$$

and this tends to $-k^{-1}$ as r tends to r_g. Thus the particle and the adventurer reach $r = r_g$ after the lapse of finite proper time for them. The singularity at $r = r_g$ is therefore not a real unphysical singularity; it is only a coordinate singularity due to the choice of coordinate systems. We shall come back to this point later. But odd things do happen at $r = r_g$. If the adventurer signals to a distant observer by sending light flashes at intervals that are precisely regular according to his (the adventurer's) proper time, the light is redshifted by a factor $g_{00}^{-1/2} = (1 - r_g/r)^{-1/2}$ as received by the distant observer ($\lambda_{receiver} = \lambda_{sender}/\sqrt{1 - r_g/r}$). This factor becomes infinite as the adventurer approaches r_g. Also, according to the distant observer, the time intervals between the received light flashes become longer and longer as the adventurer approaches r_g ($t_{receiver} = t_{sender}/\sqrt{1 - r_g/r}$). Thus, an adventurer falling radially inward appears to continue beyond the threshold at $r = r_g$, and a distant observer viewing his fall only sees him before he passes the threshold at $r = r_g$. The lack of communication with the outside world (through light or other material signals) is a basic property of black holes. The boundary $r = r_g$ is the event horizon. Just as the curved Earth leads to the existence of a horizon that limits the range of vision of a navigator on the high seas, the curved geometry of black holes also leads to the formation of an event horizon that hides its interior from the external observer of space-time. However, black holes still exert influence on their surroundings, because their gravitational effects on external bodies arise from the Schwarzschild metric for $r > r_g$.

From the above discussion we see clearly that to a distant observer the collapsing star appears to be hovering, floating just above its event horizon. It has not become a black hole. It will never. Russian scientists have coined the term "frozen star" for such an indefinitely collapsing configuration. It is understood that "frozen" means frozen in information, not in cold temperatures.

The story is quite different for the observer on the surface of the collapsing star. A co-moving observer could perform local experiments with the same outcomes as on Earth, although a strong gravitational field will be encountered (so he or she should be sufficiently small in stature so as not to be discomforted by tidal forces). However, after crossing the event horizon, he or she would experience an inexorably increasing gravitational stress that would become infinite in a finite proper time at the physical singularity.

6.2 Inside a Black Hole

We now take a brief look at how particles move inside a black hole. First, we shall see that there is no stationary particle inside a black hole. The argument runs as follows: first we recall that the world line of any particle must be time-like, i.e., the line interval ds^2 along the world line is always positive. Now, for a particle at rest, we have $dr = d\theta = d\varphi = 0$; but inside the Schwarzschild sphere or the event horizon ($r < r_g$), the coefficient $g_{00}(= 1 - r_g/r)$ of dt^2 is negative, so that

$$ds^2 = c^2 d\tau^2 = c^2(1 - r_g/r)dt^2 < 0,$$

and it is not time-like. Thus, there is no stationary particle inside a black hole. Then, how do particles move inside a black hole? Do particles move inward or outward? To answer these questions, we examine the continuity of the null cones. These are the double cones joining an event P to those neighboring events corresponding to zero separation. All possible world lines for which $d\tau^2 > 0$ (the time-like geodesics from P) all lie within the cones. The lines in one cone point into P's future, and those in the other cone point into P's past. Since the coordinates (t, r, θ, φ) are inadequate for discussing what happens at $r < r_g$, we introduce a new time coordinate that is not singular at r_g. Let us keep r, θ, φ but replace t by

$$t' = t + (2GM/c^3) \ln \left| rc^2/2GM - 1 \right| \tag{6.9}$$

from which we obtain

$$dt = dt' - (r_g/cr)(1 - r_g/r)^{-1} dr \tag{6.10}$$

and

$$dt^2 = dt'^2 - (2r_g/cr)(1 - r_g/r)^{-1} dt' dr + (r_g/cr)^2(1 - r_g/r)^{-2} dr^2.$$

The Schwarzschild metric becomes

$$ds^2 = \left(1 - \frac{r_g}{r}\right) c^2 dt'^2 - \frac{2r_g}{cr} dr dt' - \left(1 + \frac{r_g}{r}\right) dr^2 - r^2 \left(d\theta^2 + \sin^2\theta \, d\varphi^2\right).$$

$$\tag{6.11}$$

These new coordinates are Eddington-Finkelstein coordinates. Therefore, for light propagating along the radial direction ($d\theta = d\varphi = 0$), we have

$$(1 - r_g/r)dt'^2 - (2r_g/cr)dr dt' - \left(dr^2/c^2\right)(1 + r_g/r) = 0. \tag{6.12}$$

(6.12) is a quadratic equation in dt'^2 that has two roots given by

$$dt'/dr = \begin{cases} -1/c \\ c^{-1}[(1 + r_g/r)/(1 - r_g/r)] \end{cases}. \tag{6.13}$$

For $r > r_g$ the future null cones point upwards, that is, the world lines correspond to increasing coordinate time. Some lines in the future cones point inward, and some outward; that is, particles can travel either toward or away from the central mass. As we approach the black hole, the future cones lean inward until at $r < r_g$ all world lines point inward toward $r = 0$. Thus, anything (matter and radiation) inside a black hole must fall into the center $r = 0$. Figure 6.1 shows light cones in the plane of r and the new time coordinate t'. A photon starting where $r > r_g$ can

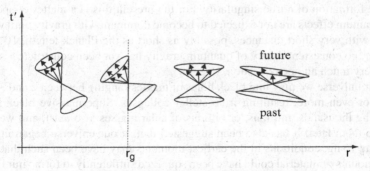

Fig. 6.1 Light cones near a black hole.

travel inward, crosses the threshold at $r = r_g$ and will carry on inward, but a photon starting where $r < r_g$ does not travel outward. Thus, an outside observer could only detect the presence of a black hole through its gravitational field.

The singularity at the Schwarzschild radius is a coordinate singularity, not an intrinsic singularity. It can be eliminated by the choice of a suitable coordinate system (see Problem 4.4). On the other hand, the singularity at $r = 0$, the center of the body, is an intrinsic one that cannot be eliminated by coordinate transformations, because the space-time curvature itself is singular there.

6.3 How a Black Hole May Form

Our discussion of the properties of a black hole would be academic unless there were reasons for believing that they might exist in nature. The possibility of their existence arises from the idea of gravitational collapse, first studied by Oppenheimer and Volkoff in 1939. Astrophysicists have calculated that the ultimate stage in the evolution of massive stars ($M > 10\,M_\odot$, where $M_\odot$ is the solar mass) would be gravitational collapse. In the process of collapse, a substantial fraction of the mass will be returned to the interstellar medium. If the mass ejected is such that what remains is in the permissible range of masses for a stable neutron star, then a pulsar will be formed. The current estimate of the permissible range of masses for stable neutron stars is from 0.3 to 2.0 $M_\odot$. The exact specification of the permissible range of masses for stable neutron stars depends on the equation of state for neutron matter. If the star ejects an amount that is either too large or too little, the residue will not be able to settle into a stable neutron star state, and the process of collapse must continue until a black hole is formed.

Black holes are very strange objects. Their most astonishing property is that, as we saw in the previous section, anything inside the Schwarzschild sphere ($r < r_g$) must fall into the center. This applies to the matter constituting the star whose collapse formed the black hole. This implies that the collapse continues until the star is a point singularity at $r = 0$. Of course, this implication holds only if general relativity remains valid in the unimaginably dense, hot conditions near the end point of collapse. It has been argued that some new force may come into play to prevent the ultimate formation of a true singularity. For the present this is a matter of speculation. Quantum effects are not expected to become dominant with gravity until we are dealing with very short distances, possibly as short as the Planck length, 10^{-35} m. However, no consistent theory of quantum gravity has yet been constructed, so this is still very much an open question.

In our universe we may find black holes of masses ranging between 2 and 3 solar masses or even more, resulting from stellar collapses. Supermassive black holes, containing thousands, millions, or billions of solar masses also exist, and we shall return to them later. It has also been suggested that, if our universe began in a hot dense Big Bang, conditions in the earliest moments may have been such that quite small amounts of material could have been squeezed sufficiently to form "mini black holes."

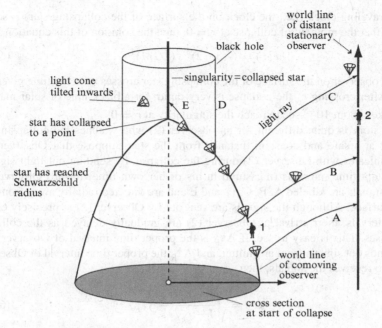

Fig. 6.2 Space-time diagram of a collapsing star.

Figure 6.2 presents the gravitational collapse in a space-time diagram, from the collapse to the development of a black hole (from bottom to top). Shown is the collapse at the center of the star, illustrated here in a circular cross section. The vertical line in the center is the world line of the star's center. As we move upward (forward in time), we see that circles of ever-decreasing radii surround the world line. These are the cross-sectional disks that collapse with time. For an observer on the surface of the collapsing star (Observer 1), nothing unusual happens as he or she crosses r_g (the Schwarzschild surface, or event horizon) at the time

$$\tau_{(r=r_g)} = -2r_g/3c. \qquad (6.14)$$

It is easy to derive the above formula. We have already worked out the necessary equation in section 6.1, which was:

$$\left(v^1\right)^2 = k^2 - 1 + r_g/r; \quad v^1 = \frac{dr}{ds}$$

Initially (i.e., before the star starts to collapse) v^1 is zero and the ratio r_g/r is negligibly small. This implies that $k^2 - 1 = 0$, so that

$$\left(\frac{dr}{ds}\right)^2 = \frac{r_g}{r} \quad \text{or} \quad \sqrt{r}\,dr = \sqrt{2GM}d\tau.$$

If the traveling clock (i.e., the clock on the surface of the collapsing star) is set to read $= 0$ at the moment of collapse at $r = 0$, then the solution of this equation is

$$r(\tau) = (-3\tau/2)^{3/2}(2GM)^{1/3} \tag{6.15}$$

and the observer on the surface of the collapsing star crosses r_g at the time given by (6.8). After crossing r_g the collapse is very rapid; for a black hole of solar mass it only takes about 10^{-5} sec to reach the singularity at $r = 0$.

The story is quite different for an observer (2) who watches the collapsing of the star at a safe and constant distance from the star. Suppose that Observer (1) communicates with Observer (2) during the collapse, by sending out light signals at constant time intervals (measured in his or her own time) toward Observer 2. Those signals are labeled A, B, C, D, and E and are directed radially away from the star's surface. Although the signals are sent out by Observer (1) at precisely equal time intervals, their arrival at Observer (2) are gradually delayed as the collapse progresses. This is easy to see. If $\Delta\tau_E$ is the proper time interval of Observer (1), who sends out signals from an emitter, and $\Delta\tau_R$ the proper time interval of Observer (2), who receives the signals, then

$$\Delta\tau_R = \sqrt{\frac{1 - r_g/r_R}{1 - r_g/r_E}} \Delta\tau_E. \tag{6.16}$$

Thus, signals A and B arrive at Observer (2) at about the same time interval with which they departed from Observer (1), and signal C arrives considerably delayed due to the increased influence of the gravitational field. Just as observer (1) crosses r_g, he or she sends signal D, which never arrives at Observer (2); it is trapped at $r = r_g$ (the vertical edge of the light cone). The last signal, E, will quickly fall into the singularity at $r = 0$. This discussion demonstrates that the collapse, when observed from a distance, will appear to decrease gradually until it stops entirely, or appears "frozen" at the $r = r_g$.

The luminosity of the collapsing star also decreases rapidly since the light will be more and more redshifted the closer to the gravitational radius r_g it is emitted. A further reduction in the luminosity results from the fact that the photons emitted at equal time intervals near Observer (1) will reach Observer (2) at ever-increasing time intervals, thus the total number of photons received per unit time is decreasing as the collapse progresses. Detailed calculations reveal that the luminosity L of the star during the last phase of the collapse near r_g diminishes exponentially:

$$L = const. \times e^{-ct/r_g}. \tag{6.17}$$

Thus, for all practical purposes, a collapsing star does appear to switch off like a light.

The description of space-time near a spherically symmetric massive object need not be in terms of the standard Schwarzschild coordinates and their corresponding line element. There are other coordinate systems available, such as the isotropic coordinates and the Kruskal coordinates. We refer interested readers to other advanced books on general relativity and cosmology for these coordinate systems.

6.4 The Kerr-Newman Black Hole

The Schwarzschild black holes we have discussed so far are of a very special kind. They are nonrotating. However, rotation is a property common to stars, planets, and galaxies alike. A rotating star possesses angular momentum, and during the collapse of such a rotating star, we may expect the angular momentum to be retained except for the part that may be radiated away in gravitational waves. If a black hole were to form as a result of the collapse of a rotating star, we would expect the black hole to be rotating at a rapid rate; after all, we know that neutron stars spin very rapidly indeed.

As a result of work carried out by R. H. Price, B. Carter, W. Israel, D. C. Robinson, and S. W. Hawking, black holes, from the point of view of the outside observer, can possess only three distinguishing characteristics: mass (M), electric charges (Q), and angular momentum (J). Roughly speaking, the reason that these properties are observed is that they are associated with long-range fields that can exert an influence at large distances. The gravitational field (associated with M and J) and the electromagnetic field (associated with Q) behave in similar fashion; they fall off with the square of distance and extend to infinite range. John A. Wheeler expresses this aspect of black holes in these oft-quoted words: "A black hole has no hair." Although black holes are mathematically very complex, they are structurally simple.

Historically, soon after Schwarzschild obtained the space-time geometry outside a spherical object of mass M, H. Reissner in 1916 and G. Nordstrom in 1918 independently solved Einstein's equations and found the space-time geometry outside a spherical object of mass M and charge Q. Then, in 1963, after a gap of 45 years, Roy P. Kerr found a solution for a black hole with mass M and angular momentum J. Some two years later E. T. Newman and others obtained solutions involving M, J, and Q, all the possible characteristics that could be possessed by black holes.

The Kerr-Newman space-time has the line element

$$d\tau^2 = -ds^2/c^2 = \frac{\Delta}{\rho^2}\left[dt - \frac{1}{c}a\sin^2\theta d\varphi\right]^2 - \frac{\sin^2\theta}{\rho^2}\left[\frac{1}{c}\left(r^2 + a^2\right)d\varphi - adt\right]^2$$
$$- \frac{\rho^2}{c^2\Delta^2}dr^2 - \frac{1}{c^2}\rho^2 d\theta^2 \tag{6.18}$$

where

$$\Delta \equiv r^2 - 2r_M r + a^2 + r_Q^2, \quad \rho \equiv r^2 + a^2\cos^2\theta, \tag{6.18a}$$

$$r_M = GM/c^2, \quad a = J/Mc, \quad r_Q = Q\left(G/c^4\right)^{1/2}. \tag{6.18b}$$

Note that r_M, a, and r_Q are mass, specific angular momentum, and charge parameters, all having the dimensions of length, a and r_Q having the same sign as J and Q respectively.

Kerr-Newman's solution has rotational symmetry about the axis $\theta = 0$; none of the metric coefficients depends on the cyclic coordinate φ. It is, moreover, *stationary*: none of the metric coefficients depends on the coordinate t that is time for an observer at infinity. For $a = 0$ and $Q = 0$, Kerr-Newman's solution reduces to Schwarzschild's solution:

$$d\tau^2 = \left(1 - \frac{2GM}{c^2 r}\right) dt^2 - \frac{1}{c^2}\left[\left(1 - \frac{2GM}{c^2 r}\right)^{-1} dr^2 + r^2 d\theta^2 + r^2 \sin^2\theta d\varphi^2\right].$$

(6.19)

For $J = 0$, but $M \neq 0$ and $Q \neq 0$, the Kerr-Newman solution reduces to the Reissner-Nordstrom solution:

$$d\tau^2 = \left(1 - \frac{2GM}{c^2 r} + \frac{GQ^2}{c^4 r^2}\right) dt^2$$

$$- \frac{1}{c^2}\left[\left(1 - \frac{2GM}{c^2 r} + \frac{GQ^2}{c^4 r^2}\right)^{-1} dr^2 + r^2 d\theta^2 + r^2 \sin^2\theta d\varphi^2\right]. \quad (6.20)$$

When $Q = 0$, the Kerr-Newman solution reduces to Kerr's solution:

$$d\tau^2 = \left(1 - \frac{2GM}{c^2 \rho^2} dt^2 - \frac{2}{c}\left(\frac{2GMr}{c^2 \rho^2}\right) a \sin^2\theta dt d\varphi\right.$$

$$- \frac{1}{c^2}\frac{\rho^2}{\Delta'} dr^2 - \frac{1}{c^2} d\theta^2 - \frac{\Lambda}{c^2 \rho^2}\sin^2\theta d\varphi^2 \qquad (6.21)$$

where

$$\Delta' = r^2 + a^2 - 2r_M; \quad \Lambda = \left(r^2 + a^2\right)^2 - a^2 \sin^2\theta. \qquad (6.21a)$$

The Kerr-Newman metric, like Schwarzschild's, has an event horizon that is spherical in shape, and its surface area A is given by the formula

$$A = 4\pi \left(r_+^2 + a^2\right) \qquad (6.22)$$

whereas its "radius" is given by

$$r_+ = \frac{1}{2}\left[r_g + \sqrt{r_g^2 + 4a^2 - 4q^2}\right] \qquad (6.23)$$

where $q = G^{1/2}Q/c^2 = r_Q$. Notice that the area A differs from the Euclidean formula, and this is due to the fact that the geometry of the black hole is non-Euclidean. For r_+ to be a real number, the quantity under the square root must be positive; that is, $r_g^2 + 4a^2 - 4q^2 > 0$. If this quantity is positive, there appears to be another horizon at

$$r_- = \frac{1}{2}\left[r_g - \sqrt{r_g^2 + 4a^2 - 4q^2}\right]. \qquad (6.24)$$

However, since $r_- < r_+$, the outside observer is concerned only with r_+.

If the quantity under the square root is negative, there will be no event horizon and we shall have a "naked singularity," i.e., a singularity that will be visible and

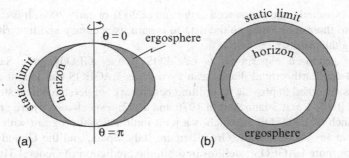

Fig. 6.3 A rotating black hole. (a) The Meridional section. (b) The Equatorial section.

communicate to the outside world. Do black holes of this type exist? Or is there a cosmic censorship that prevents naked singularity from happening? This is still an open question.

There is another surface of physical significance surrounding the event horizon. This surface is known as the *static limit* and it is not spherical. It is bun-shaped, being flattened at the poles that lie on the axis of rotation of the black hole. Figure 6.3 shows the equatorial and meridian sections of the black hole. These show that the surface of the static limit touches the event horizon at the poles. The space between the event horizon and the static limit is called the ergosphere. At latitude α, the radial coordinate for the ergosphere is given by

$$r_\alpha = \frac{1}{2}\left[r_g + \sqrt{r_g^2 - 4q^2 - 4a^2 \sin^2 \alpha} \right]. \tag{6.25}$$

Within the static limit nothing can stand still, because the space-time around a rotating object is dragged along with it. (The effect is known as the Lense-Thirring dragging of inertial frames). To understand this, consider the atmosphere dragged along with Earth's rotation. Not only bodies on Earth rotate with Earth but also bodies in the air. Birds flying up in the air and down again return to their starting places; they do not notice that Earth's surface has shifted eastward while they were in midair. That is because the atmosphere (in which the bird flies) is carried along with Earth and therefore the bird does not find any relative displacement with respect to Earth. In a manner somewhat analogous to this example, the space-time around a rotating object is carried along with it. To test this effect on the space-time around Earth, Gravity Probe B was launched on April 20, 2004, from Vandenberg Air Force Base, California. It involved putting a gyroscope in orbit around Earth. Normally, the axis of a gyroscope is fixed in space. If the general relativistic effect were present, however, the axis should precess about a fixed direction. The predicted precession rate was about $7''$ per year at a height of 800 km, not too small to be measured. This experiment was developed by Stanford University and NASA. For the past three years Gravity Probe B has circled Earth, collecting data to determine the frame-dragging effect and the other effect (the geodetic effect, the amount by which the mass of Earth warps the local space-time in which it resides). The first results confirm the two predictions of Einstein's General Relativity Theory. The final results are expected to be announced at the end of 2007 or early 2008. It is critically important to thoroughly analyze the data to ensure its accuracy and integrity prior to releasing the results.

Another proposed experiment is LAGEOS (LAser GEOdynamic Satellite), which will use Earth orbital planes as a gyroscope. LAGEOS consists of a series of satellites designed to provide an orbiting benchmark for geodynamical studies of Earth. LAGEOS 1 and 2 launched in 1976 and 1992, respectively. There are plans for the launch of LAGEOS 3, which is a joint multinational program with collaboration from France, Germany, Great Britain, Italy, Spain, and the United States. With two or more LAGEOS satellites in orbit, the prediction of General Theory of Relativity that the spin of the Earth will drag space around with it may be tested by looking for common motion of satellites in different orbits, This is referred to as the gravitational magnetic effect.

Material particles and photons can cross the static limit in either direction. Hence, unlike the event horizon, the static limit does not prevent outward leakage of information.

6.4.1 Energy Extraction from a Rotating Black Hole: The Penrose Process

The occurrence of the two separate surfaces (the event horizon and the static limit) in the Kerr-Newman geometry allows energy extraction from a black hole. This possibility derives from the fact that in the ergosphere the coordinate t, which is time-like external to the static limit, becomes space-like, and so the components of the four-momentum in the t-direction, which is the conserved energy for an observer at infinity, becomes space-like in the ergosphere. It can accordingly assume here negative values. These circumstances give rise to an unexpected energy extraction possibility. Roger Penrose in 1969 outlined a thought experiment to demonstrate this. As shown in Figure 6.4, the process involves dropping an element of matter E_0 into the ergosphere and arranging for it to break apart (in the ergosphere) into two parts in such a way that one part has a negative energy and is falling into the black hole. The black hole ends with less mass-energy, $Mc^2 - |E_1|$. For the other fragment, conservation of energy-momentum requires $E_2 = E_0 - E_1 = E_0 + |E_1|$, i.e., it is greater than the energy E_0 of the origin element. If it escapes along a geodesic, then we should have extracted the energy $|E_1|$ from the black hole.

The effect of the black hole swallowing negative energy would be to reduce its total mass-energy, and repeated application of the process could result in the extraction of a considerable fraction of the overall mass-energy. There is a limit. The negative energy particle in the ergosphere also has negative angular momentum, i.e., angular momentum opposite to that of the black hole. Thus, dropping in particles with opposite spin to the hole slows it down; when the rotation has ceased, this process can extract no further energy. If we start with a black hole spinning at the

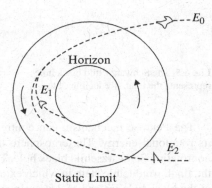

Fig. 6.4 The Penrose mechanism.   Static Limit

maximum permitted rate, by reducing its final rotation to zero we can extract 29% of the initial mass-energy of the black hole. We can get the energy extraction limits by using Hawking's area theorem, so let us introduce this area theorem first.

6.4.2 The Area Theorem

In considering the energy that could be released by interaction with black holes, Stephen Hawking discovered an important theorem in 1971. This, the *area theorem*, states that in the interactions involving black holes, the total surface area of the event horizon of a black hole can never decrease (in the absence of quantum effects); it can, at best, remain unchanged (if the conditions are stationary).

Now let us use this area theorem to estimate the energy extraction limits. For an uncharged Kerr black hole, the horizon area A is

$$A = \frac{8\pi G^2 M^2}{c^4} \left(1 + \sqrt{1 - \left(cJ/GM^2 \right)^2} \right),$$

which can be calculated from the Kerr space-time metric. This reduces to the area of a Schwarzschild black hole $A = 16\pi G^2 M^2/c^4$ that is the largest. For a maximally rotating Kerr black hole, $J = GM^2/c$, and $A = 8\pi G^2 M^2/c^4$. We now start with a maximally rotating Kerr black hole of M_i and extract some energy from it by the Penrose process to start; after the completion of the Penrose process the mass of the hole is reduced to M_f. The initial area of the black hole is $A_i = 8\pi G^2 M^2/c^4$, and the final area is $A_f \leq 16\pi G^2 M_f^2/c^4$. The equal sign applies if the black hole settles down to a Schwarzschild one after completion of the extraction process. By the area theorem we have $A_f \geq A_i$, or $16\pi G^2 M_f^2/c^4 \geq 8\pi G^2 M^2/c^4$; from this we find that $M_f^2 \geq \frac{1}{2} M^2$. Thus, at most $(1 - 1/\sqrt{2}) = 29\%$ of the initial mass-energy $M_i c^2$ can be extracted. The extracted energy is from the rotational energy of the rotating black hole. The Penrose process is not very practical, because it requires a very large break-up velocity of the particle into fragments and very accurate aim and timing. But the Penrose process is of interest for understanding nature.

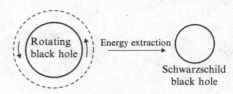

Fig. 6.5 The Schwarzschild black hole
represents the final, irreducible state.

The Penrose mechanism can continue until the black hole has given away all
its rotational energy. The ergosphere then no longer exists, and we arrive at the
nonrotating Schwarzschild black hole. Thus the Schwarzschild black hole represents
the final, irreducible state in which external processes can only increase the energy
of the black hole instead of decreasing it (Fig. 6.5).

6.4.3 Energy Extraction from Two Coalescing Black Holes

The area theorem can also be used to estimate the energy released from two coalescing black holes. Suppose two Schwarzschild black holes, each of mass M/2,
are colliding and coalescing to form a single spherical black hole of mass M'. The
surface area of the two black holes before merger is

$$dA_1 = 2 \times \left(4\pi r_g^2\right) = 8\pi \left[2G(M/2)/c^2\right]^2 = 8\pi \left(GM/c^2\right)^2. \tag{6.26}$$

In the final state, the area of the Schwarzschild black hole is

$$dA_2 = 4\pi \left(2GM'/c^2\right)^2 = 16\pi G^2 M'^2/c^4. \tag{6.27}$$

The area theorem requires that $dA_2 \geq dA_1$:

$$16\pi \left(GM'/c^2\right)^2 \geq 8\pi \left(GM/c^2\right)^2, \tag{6.28}$$

from which it follows that

$$M'^2 \geq M^2/2. \tag{6.29}$$

Hence, the maximum amount of energy that can be released in such coalescence is

$$\left(1 - 1/\sqrt{2}\right) Mc^2 = 0.293 \, Mc^2. \tag{6.30}$$

In practice, the actual amount may be much less.

The Kerr-Newman black hole appears to be the most general type of black hole.
But rotating black holes are most likely Kerr black holes. It is unlikely that black
holes can accumulate significant electrical charge, at least for long. If a black hole
were formed with, say, a strong net negative charge, it would quickly attract positive
charges in its vicinity and repel negative ones. Over a period of time the original
negative charge would be neutralized. Thus, the only practical observations of a
black hole are its mass and angular momentum. That is, a rotating black hole is
most likely a Kerr black hole, and we should set $Q = 0$ in the above discussion.

6.5 Thermodynamics of Black Holes

Hawking's area theorem opened up a new avenue in the study of black hole physics. Regarding the surface area of the event horizon of a black hole, its behavior is analogous to the behavior of a quantity known as entropy in thermodynamics, a science of the behavior of energy and information in physical systems.

The area theorem is very similar, in wording, to the second law of thermodynamics, which states that the entropy of a closed system cannot decrease. In any process that takes place, it must either increase or remain unchanged. By entropy we mean the "unavailability" of energy – energy is not available in a suitable form for useful work. Alternatively, we can regard the entropy of a system as being a measure of the disorder of that system, or of lack of information of its precise state. As entropy increases, the amount of energy available for useful work decreases, or the amount of information about the state of a system decreases.

The analogy or similarity between the behavior of entropy and the properties of event horizons led Jacob Beckenstein in 1972 to speculate that the analogy might provide a meaningful link between black-hole physics (gravitation) and thermodynamics, two apparently disparate sciences. Could a black hole possess entropy? For this idea to have merit, however, it must be possible: (a) to define precisely what is meant by the "entropy" of a black hole; and (b) to associate the concept of temperature with a black hole.

Beckenstein proposed to define the entropy of the black hole on the basis of the so-called "no-hair theorem," which was proved by Carter, Hawking, Israel, Robinson, and Price. A black hole has only three distinguishing features: mass, angular momentum, and electric charge. Beckenstein proposed and argued that the entropy of a black hole could be described in terms of the number of possible internal states that correspond to the same external appearance. In other words, the more massive the black hole, the greater the number of possible configurations that went into its formation, and the greater the loss of information. A vast amount of information is lost in the formation of a black hole. The area of the event horizon is proportional to the square of the mass of the black hole. The more massive the black hole, the greater its event horizon, and the more massive the black hole the larger the area of its event horizon. So it seems reasonable to regard the entropy of a black hole as being proportional to the area of its event horizon. It was eventually shown that the entropy of a black hole S_{bh} could be written as

$$S_{bh} = (c/4\hbar)kA \tag{6.31}$$

where A is the surface area of the event horizon, $\hbar$ is Planck's constant divided by 2π, and k is Boltzmann's constant.

With the introduction of black hole entropy, the surface of a black hole appeared to have a nonzero temperature. This was confusing because it was well known at the time that black holes absorbed all radiation that fell on it and therefore had to be at $0\,^0K$. Hawking cleared up the confusion later when he applied quantum mechanics to the region near the event horizon and discovered that black holes appear to emit particles and radiation. We shall discuss this later, but for now let us explore how we associate a temperature with a black hole.

To associate a temperature with a black hole, we can use the analogy of thermodynamics again. The temperature of a body is uniform at thermodynamic equilibrium (often called the *zeroth law of thermodynamics*). In black hole physics, a corresponding state exists for axisymmetric stationary black holes. (*Stationary* refers to a state that does not change with time and *axisymmetry* implies symmetry about some axis, i.e., the axis of rotation.) The surface of such a black hole is the same all over the event horizon. J. Bardeen developed this analogy further in 1973 as did B. Carter and S. Hawking, who showed that the surface gravity of a black hole played an analogous role to the concept of temperature in thermodynamics. The surface gravity at the event horizon of a black hole is inversely proportional to its mass, and, if the analogy is carried through, this implies also that the temperature of a black hole is inversely proportional to its mass. The less massive the black hole, the "hotter" it would be. This identification is reinforced by the following consideration. For a Kerr-Newman black hole, its "radius" and the area of its event horizon are given by

$$r_+ = \frac{1}{2}\left[r_g + \sqrt{r_g^2 - 4a^2 - 4q^2}\right] \qquad (6.32)$$

$$A = 4\pi\left(a^2 + r_+^2\right) = 2\pi\left[r_g^2 + r_g\sqrt{r_g^2 - 4q^2 - 4a^2}\right] - 4\pi q^2 \qquad (6.33)$$

respectively, where again

$$a = J/Mc; \quad q = \frac{\sqrt{G}Q}{c^2}. \qquad (6.33a)$$

Suppose we make small changes in the mass (M), angular momentum (J), and the electric charge (Q). This will result in a change of the surface area of the event horizon also. Simple calculation gives the following differential relation that connects these changes:

$$\delta(Mc^2) = \frac{\kappa c^2}{8\pi G}\delta A + \frac{AC^2}{a^2 + r_+^2}\delta J + \frac{r_+ Qc^2}{a^2 + r_+^2}\delta Q \qquad (6.34)$$

where κ is

$$\kappa = \sqrt{r_g^2 - 4q^2 - 4a^2}/2\left(a^2 + r_+^2\right). \qquad (6.34a)$$

We now compare the differential relation (34) with the thermodynamic relation:

$$\delta U = T\delta S - p\delta V \qquad (6.35)$$

where T is the temperature, S the entropy, U the internal energy of the system, and p and V are the pressure and volume of the system. This thermodynamic relation connects the increase in the internal energy of the system to the change in entropy and to the work done by (or against) the pressure. For example, if pressure puts in work and compresses the system so that the change in the volume δV is negative, this work increases the internal energy of the system. If we read these two relations in conjunction with the second law of thermodynamics and the area theorem of a black hole,

$$\delta S \geq 0, \quad \delta A \geq 0 \tag{6.36}$$

the analogy becomes clear. The area A is analogous to entropy S, the surface gravity κ is analogous to the temperature T, and the work done in changing the angular momentum or the electric charge of the black hole is analogous to the work done in changing the volume of the thermodynamic system. In each case the net result of the two relations is to change the energy of the system to the black hole or the thermodynamic system.

6.6 Quantum Mechanics of Black Holes: Hawking Radiation

The above discussion suggests that the temperature of a black hole with a finite mass is nonzero. But how can a black hole have a finite temperature? Thermal bodies with a finite temperature should emit thermal radiation in accordance with Planck's law. Yet according to the classical definition of a black hole, matter and energy could only fall into black holes; nothing could emerge from them. S. Hawking realized that this classical conclusion might not hold quantum mechanically. Therefore he investigated the quantum behavior of matter in the neighborhood of a black hole, and in 1974 he found a way out of the paradox. He discovered that black holes would appear to emit particles such as photons, electrons, and neutrinos, and that to a distant observer this radiation would have a thermal spectrum, the same kind of spectrum emitted by a black body. This startling discovery opened up the way for the establishment of links between gravitation, thermodynamics, and quantum theory.

The Hawking effect involves a quantum concept of vacuum. In classical physics, a vacuum implies absence of everything. But the quantum vacuum is a swarm of particles and antiparticles that are constantly being created and destroyed. These particles and antiparticles are considered to be virtual in the sense that they don't last long enough to be observed. This quantum concept of vacuum is related to Heisenberg's uncertainty principle. Due to the wave-particle dual nature of subatomic particles, a certain amount of uncertainty enters into the description of these particles. We cannot determine simultaneously, for example, the precise position and momentum of a subatomic particle; we can determine only the probabilities of finding particles in particular places and having particular momenta. Similarly, we cannot know precisely the exact energy of a quantum system at every moment in time. Over short time intervals, there can be great uncertainty about the amount of energy in the subatomic world. Specifically, if ΔE is the uncertainty in energy measured over a short time interval Δt, then

$$\Delta E \times \Delta t \geq \hbar. \tag{6.37}$$

We now combine this with Einstein's equation $E = mc^2$. There is nothing uncertain about c, the speed of light. Therefore any uncertainty in the energy of a physical system can be attributed to an uncertainty Δm in the mass. Thus,

$$\Delta E = c^2 \Delta m. \qquad (6.38)$$

Combining these two expressions, we obtain

$$\Delta m \times \Delta t \geq \hbar/c^2. \qquad (6.39)$$

This result is astonishing. It means that, in a very brief interval Δt of time, we cannot be sure how much matter there is in a particular location, even in a vacuum. A quantum-mechanical vacuum, in fact, is a very busy place. At any place it is possible to spontaneously create a particle-antiparticle pair. This pair can only exist for, at most, a time $\hbar/\Delta mc^2$. Before that time is up they must find each other and annihilate (Fig. 6.6a). We can therefore think of the quantum vacuum as being made up of continuously appearing and disappearing particle-antiparticle pairs. The importance of vacuum fluctuations in electromagnetic processes has long been experimentally confirmed. If electron-positron pairs are created near a real electron, the electron will attract virtual positrons and repel virtual electrons. The resulting cloud of excess positive charge surrounding the real electron cancels most of its bare charge, leaving the net small charge, $-e$, that is measured by experiments carried out at large distances from the electron. Sampled at closer distances, where the layer of shielding is partially penetrated, the measured charge would increase in magnitude. Precisely such an effect has been detected in the so-called Lamb shift of the spectral lines of the hydrogen atom.

If particle-antiparticle pairs such as $e^- - e^+$ are continuously created out of nothing as a result of fluctuation of the vacuum, then black holes can nevertheless radiate. Since energy cannot be created nor destroyed, one of the particles must have positive energy and the other one an equal amount of negative energy. They form a *virtual pair*; neither one is real in the sense that it could escape to infinity or be observed by us. However, in a strong electromagnetic field, the electron e^- and the positron e^+ may become separated by a distance of Compton wavelength λ that is of the order of the Schwarzschild radius r_g (Fig. 6.6b). Hawking has shown that

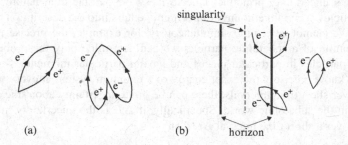

Fig. 6.6 Quantum vacuum fluctuations. (a) Virtual pairs. (b) Pair production near a black hole.

there is a small but finite probability for one of them to "tunnel" through the barrier of the quantum vacuum and escape the black hole horizon as a real particle with positive energy, leaving the negative energy particle inside the horizon of the black hole (causing the black hole to lose mass in this process). This process is called *Hawking radiation*. The rate of particle emission is as if the black hole were a hot body of temperature proportional to the surface gravity.

Quantum mechanical calculation can tell the relative chances of occurrence of these various possibilities. Hawking performed such a calculation; to his surprise and delight he discovered that the statistical effect of many such emissions leads to the emergence of particles with a thermal spectrum:

$$N(E) = \frac{8\pi}{c^3 h^3} \frac{E^2}{e^{4\pi^2 E/\kappa h} - 1} \tag{6.40}$$

where $N(E)$ is the number of particles of energy E per unit energy band per unit volume. Notice the similarity between this formula and the Planckian distribution for thermal radiation from a black body. It was this similarity that led Hawking to conclude that a black hole radiates as a black body at the temperature T

$$T = \frac{\hbar}{4\pi^2 kc}\kappa \tag{6.41}$$

where κ is the surface gravity and k is the Boltzmann constant. Note that the physical behavior of surface gravity and temperature is not merely analogous, it is identical. It is interesting to note that in the classical approximation we can set $h = 0$, then $T = 0$, and so a black hole can only absorb photons, never emit them.

Now for the Schwarzschild black hole, the surface gravity is

$$\kappa = GM/r_g^2 = c^4/4GM \tag{6.42}$$

and its temperature T can be expressed in terms of mass M

$$T = \frac{\hbar c^3}{8\pi kGM} = 6 \times 10^{-8} \frac{M_\Theta}{M} K \tag{6.43}$$

where M_Θ is the solar mass. To estimate the power radiated, we assume that the surface of the black hole radiates according to the Stefan-Boltzmann law:

$$\frac{dP}{dA} = \sigma T^4 = \frac{\pi^2}{60} \frac{k^4}{\hbar^3 c^2} T^4. \tag{6.44}$$

Multiplying by the area $A(= 4\pi r_g = 16\pi G^2 M^2/c^2)$, substituting the temperature (6.41), and noting that the power radiated corresponds to a decrease of mass M by $P = (dM/dt)c^2$, we obtain the differential equation for the mass of a radiating black hole as a function of t:

$$\frac{dM}{dt} = -\frac{\pi^3}{15360} \frac{\hbar_c^4}{G^2 M^2} \tag{6.45}$$

Using the equation $Pt = Mc^2$, we can estimate the duration of the radiation:

$$t \cong Mc^2/P \cong G^2M^3/\hbar c^4 \cong 10^{-20}M^3 \text{ sec} \qquad (6.46)$$

where M is in kg.

These results indicate that the Hawking effect for black holes with several solar masses leads to a negligible temperature and power output. The more massive the black hole, the slower the rate at which mass is being lost. A black hole with a mass comparable to that of the sun would have a temperature of about 10^{-7} K and would emit radiation at the totally negligible rate of $\sim 10^{-16}$ erg/s.

But affairs are different if the black hole has a much smaller mass. A mini black hole of the size of a proton would contain 10^{12} kg mass and have a temperature of about 10^{11} K. It would be emitting electrons, positrons, photons, neutrinos, and other kinds of particles with a power output of some 6,000 megawatts. As a black hole loses mass, its temperature increases. The hotter it becomes, the faster it radiates, and the faster it radiates, the faster it loses mass. As the mass of the black hole becomes very small the process escalates very rapidly until, in the end, the black hole radiates away the last of its mass-energy in a catastrophic explosion. At our present state of knowledge we cannot predict precisely what would occur in the final stages of explosion, but it is certain that the final explosion would result in the release of a tremendous burst of high-energy γ rays. What would be left behind after the explosion? We do not have a theory capable of explaining what happens when a black hole shrinks within the Planck radius (10^{-35} m), so the answer to this question lies in the area of speculation.

Our discussion on Hawking radiation is based on (6.45). In Hawking's original work there is a critical mass, which is the initial mass of a black hole that is at the present time undergoing the catastrophic evaporation of its remaining mass through a final burst of radiation. Hawking showed that the critical mass depends on the Hubble time (or cosmological time). At present the Hubble time is about 1.5×10^{10} yr and the critical mass is around 10^{15} g. Black holes of around 10^{15} g evaporate at a rate such that they are just now on the verge of giving up the last of their energy in a final burst. Less massive black holes will have evaporated their mass away at times closer to the origin of the universe, and more massive black holes would survive to later times. If the Hubble time were different, the critical mass would also be different. To obtain the critical mass, let us revisit (6.45):

$$\frac{dM}{dt} = -\frac{\pi^3}{15360} \frac{\hbar c^4}{G^2 M^2}.$$

This is easily integrated to give $M(t)$ in terms of M_0, the initial black hole mass. Then it follows that $M(t) = 0$ at the present time t for black holes whose initial mass M_0 is

$$M_0 = \frac{\pi}{8} \frac{1}{\sqrt[3]{10}} \left(\frac{\hbar c^4 t}{G^2 M^2}\right)^{1/3}.$$

Taking the time $t \sim 1.5 \times 10^{10}$ yr, M_0 becomes 0.8×10^{15} g. The above equation would not be the exact expression for M_0, because we omit a factor (>1) arising from the effect of back-scatted radiation and another factor (<1) due to the multiple radiation modes possible.

As shown above, a mini black hole of about 10^{16} g would be expected to evaporate over a period of about 10^{10} years, in the same order as the estimated age of the universe. Thus, we expect some such mini black holes might be exploding now, resulting in the release of tremendous bursts of high-energy gamma rays. The detection of an exploding black hole would be a discovery of utmost importance: It would (a) demonstrate the validity of the Hawking theory and of the links between gravitation, thermodynamics, and quantum theory and (b) because different theories of particle physics make quite different predictions about the properties of such an explosion, analysis of the energy emissions would provide crucial information about the nature of fundamental particles. So far no such explosions have been detected. The primordial mini black holes still remain as a theoretical possibility.

6.7 The Detection of Black Holes

We cannot observe black holes directly; we can only detect them by their interactions with other material.

6.7.1 Detection of Stellar-Mass Black Holes

Stellar-mass black holes might be detectable if they are members of binary systems.

6.7.1.1 Searches for Invisible Black Holes in Binary (or Multiple) Stellar Systems

A binary system consisting of a normal star and a black hole, circling each other at a great distance (great compared with the star's diameter) may be detected through the Doppler shift of the star's spectral lines. Aquarius is a candidate. But the problem happens to be just as difficult as it is uncertain. An "invisible" massive component need not necessarily be a black hole; the star might possibly be embedded in a dust cloud, making it invisible. However, we can detect black holes that are powerful sources of x-rays in close binary systems.

6.7.1.2 Searches for Powerful Sources of X-Rays in Binary Systems

A close binary system that consists of a black hole and a normal star can give rise to a new phenomenon. The visible component could fill its Roche lobe, and a powerful stream of gas would fall into the black hole. A Roche lobe is an imaginary surface around a star. Each star in a binary system can be pictured as being surrounded by

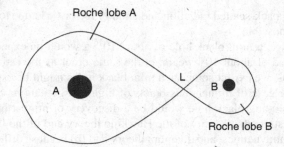

Fig. 6.7 Roche lobes of a close binary system.

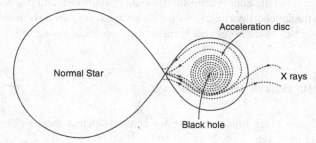

Fig. 6.8 Mass transfer from a normal star to its compact companion.

a teardrop-shaped zone of gravitational influence, called the *Roche lobe* (Fig. 6.7). Any material within the Roche lobe of a star can be considered to be part of that star. During evolution one member of the binary system can expand so that it overflows its own Roche lobe and begins to transfer matter on the other star, as shown in Figure 6.8. Since the gas stream would carry much of the angular momentum along with it, the gas would form a rapidly spinning disk around the black hole, known as the *accretion disk*. Such a laminar flow is hydrodynamically unstable, making the disc turbulent.

6.7.1.3 Turbulent Viscosity

Turbulent viscosity (and magnetic viscosity if magnetic fields are present) would cause the particles of the disc to lose angular momentum continuously, and some of them would gradually settle into the black hole. As the gas sinks into the black hole the temperature of the inner zone of the accretion disk may reach several million degrees. Such a disk could be a strong source of X-rays.

To back up our claim, let us make some rough estimates. First, to emit strongly at X-ray wavelength, say 0.3 nm (3×10^{-10} m), the temperature must be (using Wien's law) about

$$T = \left(2.9 \times 10^{-3}\right) \big/ \lambda_{\max} = \left(2.9 \times 10^{-3}\right) \big/ \left(3 \times 10^{-10}\right) = 10^7 \, K$$

The observed X-ray luminosities is from 10^{26} to 10^{31} J/s. To produce a luminosity of 10^{30} J/s at this temperature, an object that radiates like a blackbody would need a radius of

$$R = \left(\frac{L}{4\pi\sigma T^4}\right)^{1/2} = \left[\frac{10^{30}}{4\pi\left(5.7 \times 10^{-8}\right)\left(10^7\right)^4}\right] = 1.2 \times 10^4\, m \cong 10\, km.$$

That is the size of the accretion disk around a black hole. Next, we need to know the rate of mass flow onto such an object to produce the X-ray luminosity. Suppose an amount Δm falls on the surface of the object each second; the gravitational energy produced is

$$\Delta E_{gra} = GM\Delta m/R.$$

If we assume that all this energy is converted into X-rays, then $\Delta E = GM\Delta m/R = L$. From this we find Δm is given by

$$\Delta m = \frac{RL}{GM} = \frac{\left(10^4\right)\left(10^{30}\right)}{\left(6.7 \times 10^{-11}\right)\left(2 \times 10^{30}\right)} = 7.5 \times 10^{13}\, kg \cong 10^{-9}\, M/yr,$$

a rate of accretion easily obtainable in a close binary system.

The power and spectrum of the X-ray radiation from black holes look the same for neutron stars that are X-ray pulsars. The determination of the mass of the X-ray sources will give us a decisive test for distinguishing a black hole from a neutron star.

Suppose an eclipsing X-ray is discovered. Observations of the X-ray intensity produce a "light curve" (a), and observations of the radial velocity of the visible star produce a "velocity curve" (b), as shown in Figure 6.9. From Kepler's third law we have

$$(m_1 + m_2)P^2 = a^3$$

where m_1 and m_2 are the masses of the two stars (in units of the sun's mass), P is the orbital period (in years), and a is the semimajor axis of the orbit (in AU).

To apply Kepler's third law, we need to determine Period P and the semimajor axis a. We can find the period in days from either the light curve or the velocity

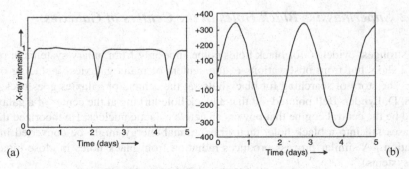

Fig. 6.9 (a) Light curve. (b) Velocity curve.

curve. To find a, we will have to use the information given on the orbital speed of the visible star, along with its period. The velocity curve shows how the observed radial velocity of the star varies as the star orbits the center of mass. The radial velocity reaches its extreme values when the star is moving directly toward or away from Earth. By finding these extreme values for radial velocity, we will find v, the orbital speed of the star. Now the star travels a distance equal to the circumference of its orbit in a time equal to its period:

$$2\pi a = vP$$

where $2\pi a$ is the circumference of the circular orbit having radius a, and vP is the distance the star travels in time P at speed v. We can solve this equation for a. Once we have both a and P, we can solve Kepler's third law for the sum of the masses of the two stars. The mass of the visible star can be determined or estimated from its spectra, using the standard tables of mass versus spectral type. The remainder is the mass of the companion object, which is the source of the X-rays. If the mass found falls into the range for neutron stars, then the X-ray is a neutron star. On the other hand, if the object is so massive, then it must be a black hole.

Close binary X-ray sources are the best suspects for containing a black hole; for example, many astronomers consider the bright X-ray source Cygnus X-1 as a possible black hole. Many investigators believe that the compact X-ray component of the Cygnus X-1 system has a mass in excess of $6\,M_\odot$, and accordingly should be a black hole.

A few other black hole candidates are known. For example, LMC X-3 (third X-ray source in the Large Magellanic Cloud) is an invisible object that, like Cygnus X-1, orbits a bright companion star. LMC X-3's visible companion seems to be distorted into the shape of an egg by the intense gravitational pull of the unseen object. Reasoning similar to that applied to Cygnus X-1 leads to the conclusion that the compact object LMC X-3 has a mass nearly $10\,M_\odot$, making it too massive to be anything but a black hole. The X-ray binary system $A0620$-00 has been found to contain an invisible compact object of mass $3.8\,M_\odot$. There are about three other known objects in or near our galaxy that may be black holes.

6.7.2 Supermassive Black Holes in the Centers of Galaxies

The strongest evidence for black holes comes not only from binary systems in our own galaxy but from observations of the centers of many galaxies, including our own. The story of searching for black holes at the centers of galaxies goes back to 1968. D. Lynden-Bell pointed out that a black hole lurking at the center of a galaxy could be the central engine that powers an active galactic nucleus. He theorized that as gases fall into a black hole, their gravitational energy might be converted into radiation. (A similar process produces radiation from black holes in close binary star systems).

To produce as much radiation as is seen from active galactic nuclei, the black hole would have to be very massive. But even a gigantic black hole would occupy a volume much smaller than our solar system – exactly what is needed to explain how active galactic nuclei can vary so widely.

In the mid-1980s, several astronomers detected a rapidly moving disk of stars surrounding the core of M31—clear evidence of an enormous mass holding them in orbit. This discovery was made by high-resolution spectroscopic observations of M31's core. By measuring the Doppler shifts of spectral lines at various locations in the core, we can determine the orbital speeds of the stars surrounding the galaxy's nucleus. It was found that the rotation curve in the galaxy's nucleus does not follow the trend set in the outer core. Rather, there are sharp peaks—one on the approaching side of the galaxy and the other on the receding side—within 5 arcsec of the galaxy's center.

The most straightforward interpretation is that the peaks are caused by the rotation of a disk of stars orbiting M31's center. One side of the disk is approaching us while the other side is receding from us. The highest observed radial velocity (at 1.1 arcsec from the galaxy's center) is 110 km/s. This is an underestimate, because unsteadiness of earth's atmosphere prohibits the detection of features smaller than about 0.5 arcsec across.

The high-speed stars orbiting close to M31's center indicate the presence of a massive central object. Calculations using Newton's form of Kepler's third law show that there must be about $10^7 \, M_\odot$ 5 pc (16 lightyears) of the galaxy's center. That much matter confined to such a small volume strongly suggests the presence of a supermassive black hole.

M32, a small satellite galaxy of M31, is an elliptical with a bright, starlike nucleus. High-resolution spectroscopy of M32 also indicates that the stars quite close to its center are orbiting the nucleus at exceptionally high speeds. These orbital motions suggest the presence of a black hole of about $3 \times 10^6 \, M_\odot$. A recent Hubble Space Telescope picture shows that the density of stars at the central region of $M32$ is more than 100 million times greater than that in our sun's neighborhood. This further supports the presence of a supermassive black hole at the center of M32.

High-resolution spectroscopy of M104 (the Sombrero galaxy), a distant galaxy 50 million light years away, reveals high-speed orbital motions around its bright, starlike nucleus. These motions suggest that $10^9 \, M_\odot$ lies within 3.5 arcsec of the galaxy's center.

Our own galaxy, the Milky Way, also harbors a monster black hole. By tracking a star near the center of our galaxy, astronomers have found the best evidence yet that a supermassive black hole lies at the Milky Way's core. The closest that the star ventures to the galaxy's center is a distance three times that between Pluto and the sun. Traveling 5,000 km/sec, the star, known as S2, takes a mere 15 years to complete one orbit of the galaxy's core. Researchers now have tracked S2 for 10 years. The star's elliptical path and high speed require the mass at the heart of the galaxy to weigh $3.7 \times 10^6 \, M$.

In total, we have about 15 mass estimates for black holes in the nuclei of nearby galaxies that are quite secure. It is believed that the majority of luminous galaxies now contain black holes. In the past, some of them outshined their host galaxies by a factor of thousands. These ancient objects are the quasars, which we see now at remote distances.

6.7.3 Intermediate-Mass Black Holes

In the year 2000, X-ray astronomers discovered a midsize black hole in M82, an unusual looking galaxy and the site of an intense and widespread burst of star formation. A Chandra image of the innermost few thousand parsecs of M82 reveals a number of bright X-ray sources close to—but not at—the center of the galaxy. Their spectra and X-ray luminosities strongly suggest that they are accreting compact objects with masses ranging from 100 to almost 1,000 times the mass of the sun. These are intermediate-mass black holes, long-sought missing links between stellar mass black holes in binaries and the supermassive black holes in the centers of galaxies. Recently, astronomers have studied the motions of stars within the globular cluster M15, the densest known in our galaxy. Using the Hubble Space Telescope, they measured a component of the velocity of individual stars orbiting within a fraction of a light year of Ml5's crowded core. Hubble's imaging spectrograph revealed that the stars close to the core move just as fast as those farther out, a strong indication that an ultradense object lurks at the globular cluster's core. They calculate that a black hole located there would have a mass of 4,000 suns.

6.8 How Do Electrical and Gravitational Fields Get Out of Black Holes?

Earlier we mentioned that a black hole possesses only three distinguishing properties: mass, electric charge, and angular momentum. These properties are preserved because they are associated with long-range fields that can exert an influence at large distances. We now take a second look at this from a quantum point of view.

The electric and magnetic fields themselves are observable, but we usually quantize the four-vector electromagnetic potentials A_μ (with $A_0 = \phi$, the electrostatic potential, A_1, A_2, and A_3 are the three components of the vector potential $\vec{A}$). To each component of the quantized four-vector potential, there is a corresponding type of photon. A complete description of the electromagnetic field can be given with photons corresponding to only three components of A_μ: two transverse and one longitudinal with respect to the direction of propagation. The electric field in electromagnetic radiation is transverse, perpendicular to the direction of the propagation of the radiation. So the electric field carried by the photon is perpendicular to the direction of propagation of the photon. In other words, the radiation field is carried by transverse photons. On the other hand, the Coulomb field is described by the longitudinally polarized photons. A transverse photon carries away energy and can be observed as a free particle; its energy E has an effective mass E/c^2. There is thus a direct interaction between a transverse photon and the gravitational field of a massive object. But the longitudinal photon does not carry energy away, and it cannot be observed as a free particle, so there is no gravitational interaction between a longitudinal photon and a massive object, and this is why a Coulomb field is able to cross the event horizon of a black hole.

As for the gravitational field, it is impossible to give a complete description of the gravitational field in terms of longitudinal and transverse gravitons alone because the Einstein field equations are nonlinear. But, within the linearized approximation, the same distinction between longitudinal and transverse photons is expected to exist between longitudinal and transverse gravitons. That is, gravitational waves are carried by transverse gravitons, so they cannot get out of a black hole. The gravitational field that reduces to the Newtonian field at large distances would be able to cross the event horizon of a black hole because it is carried by longitudinal gravitons. Thus a black hole can gravitationally attract matter and radiation outside its event horizon.

6.9 Black Holes and Particle Physics

Black hole physics forces particle physicists to reexamine the point-particle model for quarks and leptons. At extremely high energies, they cannot be point particles. Consider, for example, the collision of electrons and positrons at very high energies; their scattering behavior can be easily calculated in the standard model if they are point particles. Any deviation from the calculations indicates that electrons and positrons have substructures. So far, even at the highest energies available, there is no evidence of substructure in electrons, and their radius is less than about 10^{-17} cm. As we go to higher and higher energies, we probe smaller and smaller sizes. Is it possible that the procedure can go on to infinite energies without ever finding substructure in electrons? The answer to this seems to be "no." General relativity tells us that before we get to infinite energies, a black hole will be formed. How much energy is needed to make this happen? To answer this question, we first translate Schwarzschild's black hole condition, $2GM/c^2r > 1$, into particle physics language. We need to make three replacements:

1. Use Einstein's mass-energy formula to replace M: $M = E/c^2$.
2. Use the uncertainty principle ($\Delta p \Delta r = \hbar/2$) to replace r: $r = \hbar/2p$.
3. Use special relativity to replace p: $E = pc$ or $p = E/c$. Thus $r = \hbar c/2E$.

With these replacements, we find that in particle physics language the Schwarzschild condition becomes

$$\frac{2GM}{c^2 r} = \frac{4GE^2}{\hbar c^5} > 1$$

or

$$E > \sqrt{\hbar c^5/4G} \cong 10^{19} \text{ GeV}$$

where E is the center-of-mass energy when two point-like particles collide. At this extremely high energy, the minimum size of an electron is about 10^{-33} cm, which is in the order of Planck (or gravitational) length.

Thus, if we begin to look for the structure of the electron by colliding the electron and positron with sufficient energy, we could end with a black hole instead. This makes no sense. Quantum mechanics of point particles and the theory of classical (nonquantum) general relativity cannot be made completely consistent with one another. It tells us that something must be wrong with the theory. We know that classical concepts of space and time will fail in essential ways when we try to deal with distances smaller than the Planck length; quantum gravity is needed. On the other hand, some particle physicists deeply believe that at very high energies a theory of point particles makes no sense and we must replace them with "string" (loops of energy), "membranes" (sheets of energy), or a combination of the two. Physicists call these *membranes* "branes" for short. The theory of these objects has become known as *string theory*. When physicists try to write a complete string theory, they find that general relativity has to change, too. The string theories predict that there are at least six extra dimensions, which are all extremely small. Unwanted black holes appear in the scattering calculation until we reach an energy of around 10^{19} GeV. It is reasonable to assume that string theories become important until we work at those very high energies, and so those extra dimensions can also be very, very small – about 10^{-33} cm across. This means that the world at very small scales is not made of particles living in three spatial dimensions, but perhaps strings and branes living in nine dimensions. The string theories are the most active area of research in physics today, and physicists hope that theoretical breakthroughs will eventually allow them someday to understand how to reproduce the successes of the standard model starting from strings.

6.10 Problems

6.1. Calculate the density at which different masses (proton, Earth, sun, the Milky Way) reach their Schwarzschild radii. What conclusion can be drawn from this data?

6.2. Find the radii of circular orbits for a particle in the field of a Schwarzschild black hole. Show that the radius of the stable orbit closest to the center is given by $r = 3r_g$.

6.3. Show that the Kruskal transformation (also known as Kruskal-Szekeres transformation) given by (6.23) and (6.24) converts the Schwarzschild metric to the form given by (6.25).

6.4. Show that once a rocket ship crosses the event horizon of a Schwarzschild black hole, it will reach r = 0 in a proper time $\tau \le \pi M$, no matter how the engines are fired.

6.5. Show that the surface area of the horizon of a Kerr-Newman black hole is

$$4\pi \left(\left[M + \left(M^2 - Q^2 - a^2 \right)^{\frac{1}{2}} \right]^2 + a^2 \right)$$

6.6. Show that Kepler's law $\Omega^2 = M/r^3$ holds for circular orbits around a Schwarzschild black hole, if r is the curvature coordinate radius, and Ω is the angular frequency as measured from infinity. Derive an analogous law for equatorial orbits around a Kerr black hole of specific angular momentum a.

References

Bekerstein JD (1980) Black-hole Thermodynamics. Phys Today, 24 (January)

Berry M (1976) *Principles of Cosmology and Gravitation.* (Cambridge University Press, Cambridge, England)

Dirac PAM (1975) *General Theory of Relativity.* (John Wiley & Sons, New York)

Israel W (1987) Dark stars: the evolution of an idea. In Hawking SW, Israel W: *Three Hundred Years of Gravitation* (Cambridge University Press, 1987)

Chapter 7
Introduction to Cosmology

7.1 Introduction

Einstein's general relativity is a satisfactory theory of gravitation, and it provides a space-time structure whenever the matter distribution is given. Thus, if the average distribution of matter in the universe is put into Einstein's field equations, the average space-time structure of the whole universe may be deduced. This is a very interesting exercise, and it is part of the subject of cosmology. Cosmology is the study of the dynamical structure of the universe and seeks to answer questions regarding the origin, the evolution, and the future behavior of the universe as a whole. Historically, after establishing his General Theory of Relativity in 1916, Einstein promptly applied his theory to problems in cosmology and published his first paper on relativistic cosmology in 1917. At that time, cosmology was the only field in which the significance of general relativity could be fully manifested.

In this chapter and in the following chapters, we will be studying cosmology. Cosmologists piece together the observed information about the universe into a self-consistent theory or model that describes the nature, origin, and evolution of the universe. All model constructions are based on the following basic assumptions:

(1) The physical laws we know on Earth apply everywhere in the universe.
(2) On the large scale (100 Mpc or more), the universe is homogeneous.
(3) On the large scale, the universe looks the same in every direction (*isotropy*).

The assumptions of homogeneity and isotropy lead to the so-called *cosmological principle* that is often stated as follows:

Any observer in any place (any galaxy) sees the same general features of the universe.

This means that our local sample of the universe is no different from more remote and inaccessible regions. We will comment further on this later. The expansion of the universe (Hubble's law) and the cosmic microwave background radiation are the two basic observed pieces of information that allow us to probe the large-scale structure of the universe and construct cosmological model.

About 30 years ago it was not possible to address the central questions of cosmology with any degree of confidence. Today these questions are being explored within the framework of the Big Bang theory, which provides us with a broad outline of the evolution of the universe. In the 1970s, the Big Bang theory went through a major conceptual change. Prior to this time, cosmologists asked such questions as, "What is the average density of matter in the universe?" "How rapidly is the universe expanding?" At this time cosmologists began seriously asking question like, "Why does matter exist at all, and where did it come from? Why is the universe as homogeneous as it is over such vast distances? Why is the cosmic density of matter such that the energy of expansion of the universe is almost balanced by its energy of gravitational attraction?" In other words, the investigations became more fundamental. "Why?" was added to "What?," "How?," and "Where?".

7.2 The Development of Western Cosmological Concepts

7.2.1 Ancient Greece

Every culture has had its cosmology, its story of how the universe came into being, what it is made of, and where it is going. The mythological stories can be traced to the earliest writings of the Babylonian, Egyptian, Greek, and Chinese civilizations. The transition from mythology to the birth of scientific inquiry occurred rather abruptly in the middle of the sixth-century B.C. on the shores of Asia Minor. The earliest surviving attempt at a rational cosmology was probably that of Pythagoras, who taught that

(1) Earth is round and rotates on its axis.
(2) The sun, moon, stars, and planets revolve on concentric spheres around a central fire. The "fixed" stars form the outermost sphere.
(3) The motion of the celestial bodies produces the harmony of the musical scale.

Although Pythagorean philosophy prepared the way for a heliocentric cosmology and persisted for several centuries, its emphasis on celestial harmony based on musical scale made it eventually obsolete.

The ideas of Plato and Aristotle appeared around fourth century B.C. Plato held that the circle was a perfect form, and therefore the celestial motions had to be in circles, since the universe was created by a perfect being, God. Plato also advocated the idea of daily rotation of the heavens around a spherical, immovable Earth. The planets moved in circular orbits at different rates, with Mercury and Venus moving from west to east, but the other heavenly bodies moved from east to west. Plato took little interest in observation of the heavenly motions. He did not notice, for example, that the apparent westward motions (the retrograde motion) of Mercury and Venus occur only during part of their orbits.

Eudoxus, a younger contemporary of Plato's, made a serious attempt to account for the retrograde motions of the planets. His work constituted the first really scientific astronomy – not just philosophical speculations without any observational basis. Aristotle further modified Eudoxus' scheme. Together, they prepared the way for a geocentric cosmology.

In about 280 B.C., Aristarchus offered a model that the planets, including Earth, revolved in circular orbits around the sun, which is a vastly simpler model than that of Eudoxus and Aristotle. However his views were eclipsed by Aristotle's fame. The other Greek philosophers of his time were reluctant to explore the implications of the theory of planetary motions implicit in Aristarchus's heliocentric theory. Some five centuries later, the Greek philosopher Ptolemy, who lived in Alexandria during the second century A.D., introduced a geocentric cosmology, which was adopted later by the Roman Catholic Church as an article of faith. Thus the Ptolemaic theory was not seriously challenged for 1,400 years. The destruction brought about by the barbarian hordes in the sixth century devastated the Roman Empire, and the fruits of Greek learning were swept aside. The dark Middle Ages commenced, and scientific progress was set back a thousand years or more.

7.2.2 *The Renaissance of Cosmology*

During the 13th century the works of ancient Greek philosophers were translated back from Arabic translations. The Ptolemaic system became widely known in the course of the following two centuries, and was not seriously questioned until Nicolaus Copernicus (1473–1543) reexamined it in the early 1500s.

Copernicus introduced a heliocentric system. He showed that the motion of the planets around the sun, with the moon orbiting around a rotating Earth, provided a far simpler and more elegant explanation of planetary motion. Copernicus was primarily concerned with planets and did not take the logical step of recognizing that the stars are scattered throughout space. Thomas Digges, an Englishman, took that step in 1576.

The next great advance came as a result of serious observations of planets by Tycho Brahe (1546–1601). Brahe noticed that the observation of stellar parallax should provide a clear test for the geocentric and heliocentric systems, and he devised and performed numerous measurements on stellar parallax without success. As a result, he advocated a geocentric solar system, in which the planets revolved around the sun, which itself orbited the stationary Earth. His compromised model failed to account for the most obvious aspects of the motions of the planets. However, he is owed recognition for his ingenuity, precision, and great faith in observational data. His other contribution to cosmology was to demonstrate that comets were much more distant than the moon and had highly elongated orbits. This discovery discredited the Aristotelian notion of heavenly spheres that were fixed, permanent, and solid.

Tycho's data passed to his assistant Johannes Kepler (1571–1630), who finally formulated the three laws of planetary motion. Kepler made a lasting contribution of great significance. He could achieve such great success because he was able to break away from preconceived notions and discard circular orbits.

Tycho Brahe was a great observational astronomer; Galileo Galilei (1564–1642) was even greater. He pioneered scientific advancement by innovating systematic methods of observation and experiment. He used the newly developed primitive telescope to discover the phases of Venus, which are much like our moon. This shows that Copernicus was right – Venus revolves around the sun. The discovery of four large satellites of Jupiter showed that Earth is clearly not the center of all motion in the cosmos. Galileo himself did not contribute significantly to cosmological theory, but his discoveries made a path for others to follow. After Galileo, scientists relied more and more on evidence, observation, and measurement.

Kepler and Galileo were unable to explain why the planets move around the sun in elliptical orbits and what keeps the solar system together. Isaac Newton (1643–1727) provided the underlying theory, the law of gravity. He used it to explain Kepler's laws of planetary motion. William Herschel found that binary stars in orbit around one another obey Newton's law of gravity. This discovery demonstrates the universality of Newton's law of gravity. Cosmology received an enormous boost when Herschel observed nebulae through his 72-inch reflecting telescope. He considered these nebulae to be "island universes" of stars. Thomas Wright and I. Kant had previously speculated about such nebulae. Herschel's observations not only verified their existence but also established extragalactic astronomy as a new frontier.

7.2.3 Newton and the Infinite Universe

The ancients never contemplated the possibility of an infinite universe. Both geocentric and heliocentric systems regarded the universe as having a finite space with the visible stars fixed to an outer most sphere around Earth or the sun. It was Thomas Digges who introduced the concept of infinity to the modem picture of the universe. he dispersed the stars in the geocentric and heliocentric systems, star sphere, into an endless infinity of space stretching out across the universe. He also acknowledged the need to explain why, in an infinite universe, the sky should be dark at night.

In 1610 Kepler argued that the darkness of the night sky directly conflicted with the idea of an infinite universe filled with bright stars. This led him to believe that the universe was finite in extent. But Newton firmly believed that the universe was infinite in extent, with stars scattered more or less randomly throughout space. He argued that if the universe were finite, or if stars were grouped in one part of the universe, the gravitational forces would soon cause all stars to collapse into a huge clump at the center, but an infinite universe has no center so it can't collapse. After Newton, the concept of an infinite steady universe became firmly established. Newton ignored the dark night puzzle. We know today that Newton's argument for an infinite, static universe does not hold water.

7.2.4 Newton's Law of Gravity and a Nonstationary Universe

Actually Newton's own law of gravity predicts a nonstationary universe. To this purpose, let us first introduce an important property of Newton's theory of gravity: that a hollow spherically symmetric shell of matter does not create any gravitational field in its interior.

Consider a thin spherical shell of matter as shown in Figure 7.1. We are going to compare gravity forces that pull a particle of mass m (located at an arbitrary point inside the shell) in two opposite directions, a and b. The direction of the line ab, passing through m, is supposed to be arbitrary, too. The forces of gravitational attraction are created by the matter within the two surface elements cut out from the shell by two narrow cones with equal vertex angles. The areas of the surface elements cut by these cones are proportional to the squares of the cone heights. Namely, the ratio of the area a of element a to the area S_b of element b is equal to the ratio of the squares of the distance r_a and r_b from m to the shell surface along the line ab:

$$S_a/S_b = r_a^2/r_b^2. \tag{7.1}$$

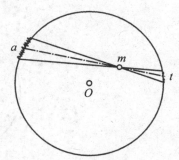

Fig. 7.1 A hollow spherical symmetric shell of matter has no gravitational field in its interior.

If the mass is to be evenly distributed over the shell surface, we arrive at the same ratio for the masses of the surface elements:

$$M_a/M_b = r_a^2/r_b^2. \tag{7.2}$$

Now we can calculate the ratio of the forces with which surface elements attract the particle. According to Newton's law the expressions for these forces are

$$F_a = GM_a m/r_a^2, \quad F_b = GM_b m/r_b^2$$

Their ratio is given by

$$F_a/F_b = M_a r_b^2/M_b r_a^2. \tag{7.3}$$

Substituting for M_a/M_b in (7.3) its value from (7.2), we finally get

$$F_a/F_b = 1, \quad F_a = F_b.$$

Hence, the two forces are equal in magnitude and act in opposite directions, thus canceling each other out. The argument can be repeated for any other direction. As a result, all forces pulling m in opposite directions cancel one another out, and the resultant force is exactly zero. The location of particle m was arbitrary. Hence, there truly are no gravitational forces inside any spherical shell.

Now let us turn to the effect of gravitational forces in the universe. The distribution of matter in the universe is homogeneous on a large scale. Since we discuss large scales only, we assume the matter to be uniformly distributed in space.

Let us single out of this uniform background an imaginary sphere of an arbitrary radius with the center at an arbitrary point, as depicted in Figure 7.2. Consider first the gravitational forces that the matter inside the sphere exerts on the bodies at its surface, ignoring for a moment the effect of matter outside the selected sphere. Let the radius of the sphere not be too large, so that the gravitational field generated by its interior is relatively weak and the Newtonian theory of gravity applies. Then, the galaxies at the surface of the sphere are attracted to its center by forces that are directly proportional to the mass M of the sphere, and inversely proportional to the square of its radius R.

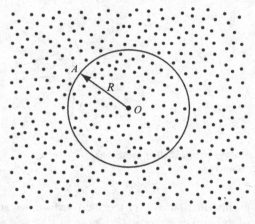

Fig. 7.2 The gravitational force on body (A) is determined only by matter inside the sphere.

Next, consider the gravitational effect of all the remaining material in the universe, which lies outside the sphere under consideration. All this matter can be thought of as a sequence of concentric spherical shells with increasing radii, surrounding our selected sphere. But, as we have already shown, spherically symmetric layers of material create no gravitational forces in their interiors. As a result, all the spherical shells (i.e., all the remaining material of the universe) add nothing to the net force attracting some galaxy A at the surface of the sphere toward its center O.

So, we can calculate the acceleration of one galaxy A with respect to another galaxy O. We have associated O with the center of the sphere, while A is at a distance R from it. The acceleration to be found is due to the gravitational attraction of matter inside the sphere of radius R only. According to Newton's law it is given by

$$a = -GM/R^2 \tag{7.4}$$

The negative sign in (7.4) reflects the fact that the acceleration corresponds to attraction rather than to repulsion.

Thus, any two galaxies in a homogeneous universe separated by a distance R experience a relative acceleration a as given by (7.4). This means that the universe cannot be stationary. Indeed, even if one assumes that at some instant all the galaxies are at rest and the matter density in the universe does not change, the very next moment the galaxies would acquire some speed due to mutual gravitational attraction resulting in the relative acceleration represented by (7.4). In other words, the galaxies could remain motionless with respect to each other only for a brief instant. In the general case they must move – either approach each other or recede from each other. The radius of the sphere R (see Fig. 7.2) must change with time, and so must the density of matter in the universe.

The universe must be nonstationary because gravity acts in it; that is the basic conclusion from the theory. A.A. Friedmann first reached this conclusion, in the framework of Einstein's relativistic gravitational theory, from 1922 to 1924. Some years later, in the mid-1930s, E.A. Milne and W.H. McCrea pointed out that the nonstationary behavior of the universe can be also derived from Newtonian theory as outlined above.

Furthermore, Newton's law of gravity also predicts that an infinite steady universe is an empty universe; that is, it contains no matter at all. To see this, let us first rewrite Newton's law of gravity

$$F = GMm/r^2$$

in terms of field strength. The strength of the field, Γ, produced by M at any point in space is defined as the gravitational force that a unit mass would experience if placed at that point. Thus the force on m, when it is placed in the gravitational field of M, is the product of the field strength Γ and the mass m

$$F = m\Gamma.$$

Thus,

$$\Gamma = F/m = GM/r^2.$$

Or

$$\Gamma = \frac{GM}{r^2} = \frac{4\pi GM}{4\pi r^2} = \frac{4\pi GM}{A},$$

where $A = 4\pi r^2$ is the surface area of a sphere of radius r centered at M. From the last equation we obtain

$$A\Gamma = 4\pi GM.$$

In an infinite universe this small volume V will be equally attracted in all directions, and so on the average Γ vanishes. Then $M = 0$, as $4\pi G$ cannot be zero. That is, the small spherical volume contains no matter. Since the small volume V is arbitrary, it could be anywhere in the universe, and so $M = 0$ everywhere in the universe. That is, an infinite steady universe contains no matter and is an empty universe. Obviously this is not the case with the real universe.

7.2.5 Olbers' Paradox

The dark night puzzle, used by Kepler in 1610 to advocate the idea of a finite universe, is known today as Olbers' paradox. We know that the sky is dark at night. Digges recognized in 1517 that in a static, infinite universe the night sky should not be dark. In 1826 Heinrich Olbers, a physician and an amateur astronomer, detailed his discussion of the dark night puzzle in a paper. He investigated the dark night puzzle based on what were then very reasonable assumptions:

(1) The stars are evenly distributed throughout infinite space, and their absolute brightness is the same everywhere and at all times.
(2) The stars are at rest, except for local random motions.
(3) The universe does not change with time.

With these assumptions, Olbers found a very strange result: the sky should be everywhere as bright as it would be at the surface of the sun. To see this strange result, consider a thin spherical shell of thickness t, the center at the observer O (Earth), with an inner radius r; the number of stars, N, in the shell is given by

$$N = 4\pi r^2 t n$$

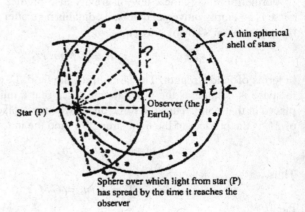

Star (P)

Observer (the Earth)

A thin spherical shell of stars

Sphere over which light from star (P) has spread by the time it reaches the observer

Fig. 7.3 Olbers's paradox.

where $4\pi r^2 t$ is the volume of the spherical shell, and n is the number of stars per unit volume. If l is the amount of light emitted by an individual star, then the amount of light emitted by the stars in the shell is given by

$$L = 4\pi r^2 t n l.$$

How much light from the shell will reach the observer O at the center of the spherical shell? The amount of light reaching O from one star in the shell is given by $l/(4\pi r^2)$, so the amount of light the observer O receives from all the stars in the shell is

$$(4\pi r^2 t n) \times \frac{l}{4\pi r^2} = t n l.$$

We see that the radius r cancels out, and the amount of light reaching the observer O at the center from the shell is proportional simply to its thickness t. The same result will apply to any shell centered on O, whatever its radius. Since there are an infinite number of such shells in an infinite universe, the total light to reach O is infinite! This calculation ignores that fact that some light is intercepted on the way by stars between the emitting star and the observer. But even if account is taken of this, the result is the same. The sky would not the difference between day and night. Clearly this is not so. This dilemma is known as Olbers' paradox.

Olbers was naturally very puzzled by this absurd result derived from such plausible assumptions. In his time it did not seem possible for any of these to be wrong. He concluded there must be a lot of dust between the stars and Earth, which absorbs the greater part of the light. Today we know that this explanation is wrong: the dust would eventually become so hot that it would emit as much light as it received. Hence it would have no shielding effect.

Olbers' paradox tells us there is something very wrong with the idea of an infinite, static universe. The expansion of the universe has resolved Olbers' paradox. There are two aspects of the expansion of the universe that help: redshift and the finite age of the universe.

The expansion of the universe implies that the universe has a finite age. So the stars beyond some finite distance, known as the horizon distance, are invisible to us because their light has not had enough time to reach us.

Redshift of starlight also contributes significantly to the resolution of Olbers' paradox. According to the new view of Big Bang, the expansion of the universe is the expansion of space. As a photon travels through the expanding space, its wave length becomes stretched. That is, it is redshifted. Because the velocity of light is finite ($c = 3 \times 10^{10}$ cm/ sec), the farther we look into space, the farther we go back in time, and the light is more redshifted. Cosmologists often call the travel time of light "the lookback time." The redshift has a doubly weakening effect on light. First, since the wavelength of the incoming waves is increased, their frequency is reduced ($f = c/\lambda$); this diminishes their energy, according to Planck's formula

$$E = hf, \quad h = \text{Planck constant}$$

Second, the lowering of the frequency means that not fewer photons (particles of light) arrive in one second, so that the energy received is still further reduced.

It is amazing that great scientists such as Newton and Einstein ignored Olbers' paradox and missed the opportunity to realize or conclude that the observable universe has a finite size and age.

7.3 The Discovery of the Expansion of the Universe

The development of spectroscopy led to many surprising discoveries in astronomy. Over the 20 years from 1912 to 1932, Vesto Slipher managed to obtain the spectra of the light from some 40 relatively nearby galaxies. Slipher's observational

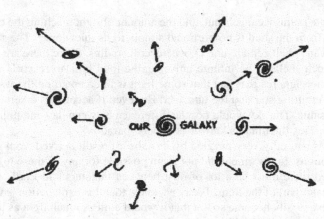

Fig. 7.4 The recession pattern of galaxies as seen from our galaxy.

achievement was exceptional; galactic spectra are very faint and complex, since the light originates from millions of individual stars, each with its own motion within the galaxy. Through the Doppler effect this spread in velocities leads to a spread in the frequencies and wavelengths of spectral lines. Nevertheless, Slipher discovered that the spectra of most galaxies showed an overall redshift, which implied that the galaxies were receding from us at quite significant speeds; the fainter the galaxy, the great its redshift. We cannot learn too much from these redshifts alone. The difficulty was in knowing whether Slipher was looking at bright objects a long way off or at dim ones nearby. Thus one of the most urgent issues in cosmology was the distance scale. The pioneering work of Edwin Hubble resolved this issue. He used Cepheid variables as distance indicators and demonstrated that the redshift was directly proportional to the distance of the galaxy; the greater the distance to a galaxy, the greater its apparent recession velocity (Figure 7.4). This proportionality relation between velocity and galactic distance is known as Hubble's law (or the law of redshift):

$$V = H_0 r$$

where V is the speed of recession of a galaxy, r is its distance, and H_0 is a proportionality constant, called Hubble's constant today in honor of Hubble. The implication of Hubble's result was revolutionary: the universe is expanding! Figure 7.5 is a plot of the recession velocity versus apparent distances for a group of spiral galaxies used by astronomers for calibrating distances. The slope of the line is H_0. And Figure 7.6 shows the redshifted spectra of three galaxies whose distance distances from us range from 72 to 3,800 million light years.

Slipher's original finding that more galaxies move away from us than toward us seemed odd at first. Now it is obvious, because nearby systems possess local peculiar motions that can be greater than the redshifted expansion velocity.

The Hubble constant is one of the most important numbers in all astronomy. It expresses the rate at which the universe is expanding and gives the age of the universe. The measurement of the Hubble constant is difficult, and its stated value

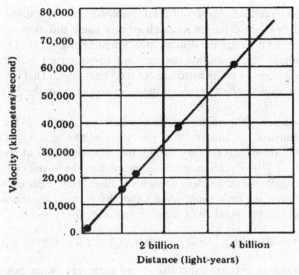

Fig. 7.5 Radial velocity-distance relation for galaxies according to Hubble.

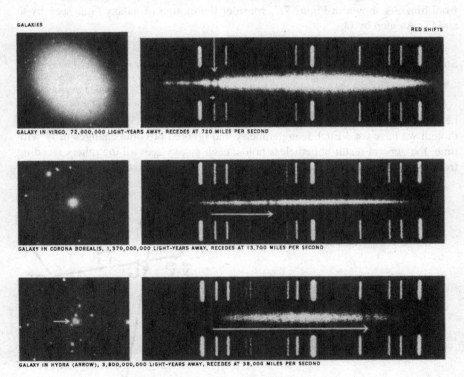

GALAXIES   RED SHIFTS

GALAXY IN VIRGO, 72,000,000 LIGHT-YEARS AWAY, RECEDES AT 720 MILES PER SECOND

GALAXY IN CORONA BOREALIS, 1,370,000,000 LIGHT-YEARS AWAY, RECEDES AT 13,700 MILES PER SECOND

GALAXY IN HYDRA (ARROW), 3,800,000,000 LIGHT-YEARS AWAY, RECEDES AT 38,000 MILES PER SECOND

Fig. 7.6 Three galaxies and their spectra.

is constantly being updated. As of mid-2001, published measurements of H_0 over a five-year period by many different research groups, using different sets of galaxies and a wide variety of distance-measurement techniques, give results between 50 and 80 km/s/Mpc, where Mpc is million parsecs, and 1 parsec (pc) = 3.26 light years. The number used often by many astronomers is 65 km/s/Mpc. That is to say that for each million parsecs distance to a galaxy, the recession speed of the galaxy increases by 65 km/s.

Hubble's law asserts that the universe is in uniform expansion. What do we mean by uniform expansion? To answer this, we consider two galaxies, G and G', at positions $\vec{r}$ and $\vec{r}'$ from us (O). They form a triangle with sides of length $r = |\vec{r}|$, $r' = |\vec{r}'|$, and $\bar{r} = |\vec{r} - \vec{r}'|$. Homogeneous means that the shape of the triangle is preserved as the galaxies move away from each other. This requires that each side of the triangle increases by a same scale factor $R(t)$, equal to one at the present moment ($t = t_0$) and independent of location and direction:

$$r(t) = R(t)r(t_0), \quad r'(t) = R(t)r'(t_0), \quad \bar{r}(t) = R(t)\bar{r}(t_0)$$

Since the universe is in uniform expansion, there are no privileged positions, and so an observer moving together with any galaxy sees the surrounding galaxies receding from him. As shown in Figure 7.7, consider the motion of galaxy G as seen by O and G'. As seen by O,

$$\vec{v} = H\vec{r}, \vec{v}' = H\vec{r}'$$

So $\vec{v} - \vec{v}'$, the velocity of G relative to G', can be expressed as

$$\vec{v} - \vec{v}' = H\vec{r} - H\vec{r}' = H(\vec{r} - \vec{r}')$$

i.e., G' also sees G, and therefore all the other galaxies, receding from itself. Although we have used Euclidean geometry and ignored possible changes in H with time, the general result nonetheless holds: each galaxy sees all the others receding from itself, and we not at the center of this expanding universe.

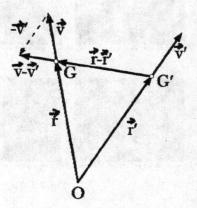

Fig. 7.7 Expansion as seen by O and G'.

7.4 The Big Bang

Since the universe is expanding, there must have been a time in the very distant past when everything in the universe – matter and radiation alike – was concentrated in a state of infinite density at a point. Presumably this exploded and initiated the expansion of the universe. This scenario is called the Big Bang. In other words, the Big Bang was an explosion of space at the beginning of time. As time elapses, space itself expands. The expansion of the universe is the expansion of space. The universe has no center and no edge. If you take a balloon, blow it up, and mark a number of small dots on the surface, then blow it up some more. You will see all the dots move away from each other. This is rather like the expansion of the universe: the expansion of the universe is not by the galaxies moving through space, but rather it is that the space between the galaxies is expanding. The three "spatial" dimensions of our universe can be thought of as the two dimensions on the surface of the balloon. A creature can only crawl around the surface, never finding an edge or the center.

It was many years after Hubble's discovery that a reliable Big Bang theory was developed. The main reason for this was that the physics of the processes going on at the high energies involved at the early stages of the universe was not known to Hubble or his contemporaries.

How long ago the Big Bang took place depends on the value of the Hubble constant (and the models of the Big Bang). We can make a crude estimate by asking how long a distant galaxy has been traveling, assuming that it has had constant speed since the moment of the Big Bang. (It is expected that the rate of expansion reduces as time goes on because the expansion is resisted by the gravitational attraction of the matter in the universe.) Using Hubble's law, we see that this time is given by

$$t = \frac{r}{V} = \frac{1}{H_0} = \frac{3.09 \times 10^{22} m}{75 \times 10^3 m/s} = 4.1 \times 10^{17} s = 1.3 \times 10^{10} \, \text{yr}.$$

This is a rough estimate, but it is in the right order of magnitude.

The expansion of the universe gives the redshift of light from remote sources a proper interpretation. As the universe expands, the wavelengths of all photons expand in proportion to the scale increase. Consider radiation emitted at wavelength λ_1 at time t_1 from a galaxy G and detected at wavelength λ_0 at time t_0 by a co-moving observer O. The scale of the universe grows from R_1 to R_0 during the interval that radiation from G travels to observer O. The wavelength of radiation increases during transit by a factor R_0/R_1

$$\frac{\lambda_0}{\lambda_1} = \frac{R(t_0)}{R(t_1)},$$

but

$$\frac{\lambda_0}{\lambda_1} = 1 + z$$

and so

$$1 + z = \frac{R(t_0)}{R(t_1)}.$$

The redshift caused by the expansion of the universe is properly called a *cosmological redshift*.

If G is not very far away from O, we can use the rate of change of R at t_0 to estimate the value of R at t_1:

$$R(t_1) \approx R(t_0) - \dot{R}(t_0)(t_0 - t_1)$$

from which we have

$$\frac{R(t_0)}{R(t_1)} = 1 + \frac{\dot{R}(t_0)}{R(t_1)}(t_0 - t_1)$$

and so

$$z = \frac{\dot{R}(t_0)}{R(t_1)}(t_0 - t_1).$$

On the other hand, we may compute the distance of the galaxy, D, from the time the radiation has taken to traverse it:

$$D = c(t_0 - t_1).$$

Eliminating $(t_0 - t_1)$ from the last two equations, we obtain

$$z = \left[\dot{R}(t_0)/R(t_1) \right] [D/c] = (H/c)D,$$

which is Hubble's law, provided we identify

$$H = \dot{R}(t_0)/R(t_1).$$

7.5 The Microwave Background Radiation

The basic observational evidence of the expanding universe is that light from distant galaxies is shifted in wavelength toward the red end of the spectrum. What is the evidence of a hot Big Bang? It is the microwave background radiation, a small remnant of radiation left over from the hot Big Bang. As we shall see, this microwave background radiation is crucial to making detailed predictions in Big Bang cosmology.

In the late 1940s George Gamow and colleagues pointed out that if the universe began with a hot Big Bang, as they thought likely, the blackbody radiation emitted at that time should still be present. The universe has expanded so much since the Big Bang that all short-wavelength photons today have wavelengths that are so stretched that they have become long-wavelength, low-energy photons. This cosmic radiation field would look like the radiation emitted by a blackbody at a very low temperature. Gamow had predicted a current temperature of about 5 K. At the time Gamow

Fig. 7.8 Robert Wilson and Arno Penzia (left to right) in front of big horn antenna.

made this prediction, equipment capable of detecting such radiation was not available, so nothing came of the suggestion that the radiation might still be bouncing around space. In the early 1960s Robert Dicke at Princeton University had arrived independently at a prediction of such radiation at 10 K by a different route. Dicke and his colleagues began designing an antenna to detect this microwave radiation.

Meanwhile, just a few miles from Princeton University, Arno Penzias and Robert Wilson of Bell Laboratories (Fig. 7.8) had modified a ground-based radiometer that had been used to detecting signals from Echo satellites, into a low-noise radio antenna of 7.35 cm long. They were bothered by an excess of noise that they could pinpoint – it did not vary with the time of day or the season. Assuming it was due to radiation in thermal equilibrium, they calculated the antenna temperature of this blackbody radiation. They first applied Wein's displacement formula, $\lambda_{peak} = 0.51 \, cm/T(K)$, with $\lambda_{peak} = 7.35$ cm, and found that $T = 0.51/7.35 = 0.07 \, K$. This was unreasonably low, so they assumed they were below the peak and on the low frequency or long-wavelength tail of the blackbody spectrum. With this last assumption and the measured value of the energy density per unit frequency, they found the predicted antenna temperature to be $\sim 3 \, K$, and became convinced that the noise was not caused by their instrument. They conjectured that it might be extraterrestrial. After learning about the work of Dicke and his colleagues at Princeton University, Penzias and Wilson came to realize that they had discovered the background radiation left over from the hot Big Bang.

Since those pioneering days, scientists have made many measurements of the intensity of the cosmic background radiation at a variety of wavelengths. The most accurate measurements come from the Cosmic Background Explorer satellite, which was placed in orbit around Earth in 1989 (Fig. 7.9). Data from COBE's spectrometer (Fig. 7.10) demonstrate that this ancient radiation has the spectrum of a blackbody with a temperature of 2.73 K. This radiation field, which fills all of space, is commonly called the *cosmic microwave background*.

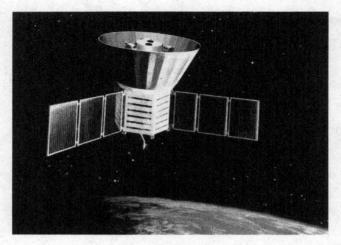

Fig. 7.9 COBE Explorer.

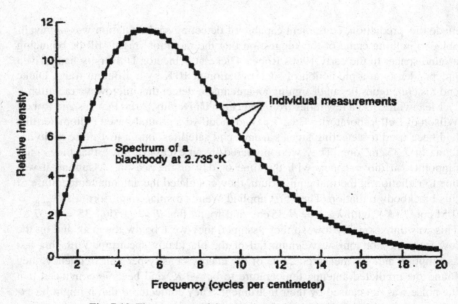

Fig. 7.10 The spectrum of the cosmic microwave background.

We will learn later that this cosmic background radiation was released about half a million years after the expansion began, at a time when hydrogen atoms had cooled to 3000 K, the temperature at which atoms are stable. The universe was then a neutral gas of atoms, and the electromagnetic radiation present at that time could travel without being absorbed. Following that period, very little additional electromagnetic radiation has been formed, since neutral atoms do not radiate nearly as readily as charged particles. The spectrum of the microwave radiation now observed, therefore, reflects the temperature $t = 10^5$ years. However, it is extremely redshifted, and so we now measure 2.73 K as its temperature, rather than the few thousand Kelvins characteristic of dissociation of atoms.

Some of you may wonder how the cosmic background radiation has the spectrum of a blackbody while the space is expanding. So before we continue further, let us digress a moment to take a look at the effect of expansion on the spectrum of the cosmic radiation.

We may visualize the effect of the adiabatic expansion of the universe in the following way. Imagine a box made of perfectly reflecting mirrors and blackbody radiation from a hot source is directed to the box. Next, the box is closed so that there can be no leakage of the radiation. The radiation will be trapped and bounce back and forth between the walls indefinitely. Now let the walls of the box slide outward so that the volume of the box increases. As radiation strikes the moving walls, it undergoes Doppler shifts to the red, so the wavelengths of the entire radiation increase. But the wavelengths increase in such a way that the distribution among them still corresponds to the radiation curve for a blackbody. The effect of the moving walls, as the box becomes larger, is to change the radiation from that corresponding one temperature to that for a blackbody at a lower temperature.

An important feature of the cosmic microwave background radiation is its intensity. At any given wavelength, the cosmic background radiation is extremely isotropic on the small scale. In directions that differ by only a few minutes of arc, any fluctuations in its intensity is less than 1 part in 10,000. On the other hand, a large-scale anisotropy in the cosmic background radiation has now been established, in the sense that it is slightly hotter in one direction than in the opposite direction in the sky. This is due to our own motion through space. If we approach a blackbody, its radiation is Doppler-shifted to shorter wavelengths and resembles that from a slightly hotter blackbody. When we move away from it, the radiation appears like that from a slightly cooler blackbody. This effect has been observed in the microwave background.

The measurements of this relative speed are very difficult because the difference in intensity is very tiny compared with the radiation from Earth's own atmosphere. Hence the measurements must be made from high-flying balloons, aircraft, or spacecraft. The data indicate that our Galaxy and the Local Group is moving at a speed of about 600 km/s with respect to the microwave background (or with respect to the uniform expansion of the universe as a whole), toward the general direction of the Virgo and Hydra cluster. This can be thought of as a peculiar motion of the Local Group, superimposed on the general expansion.

The uniformity of the radiation tells us that at an age of less than a million years the universe was extremely uniform in density. But at least some density variations had to be present to allow matter to gravitationally clump up to form galaxies and superclusters of galaxies. The isotropy of the microwave background radiation, therefore, puts interesting constraints on theories of supercluster, cluster, and galaxy formation.

COBE's Differential Microwave Radiometer, a set of very sensitive and stable radio receivers, was designed to analyze this problem by mapping the cosmic background far more precisely than is possible from Earth's surface. Even above the atmosphere, the cosmic background fluctuations are swamped by radiation fluctuations due to foreground stars and dust clouds in the Milky Way and other galaxies. Such interference must be identified and subtracted from the measured signal. Hundreds of millions of observations were processed in this fashion to produce a single map. This map was then analyzed statistically and revealed the presence of fluctuations of $(30 \pm 5) \times 10^{-3}$ K in the temperature of the background radiation. Indeed, COBE had detected the non-uniformity in the microwave background, amounting to about 30 millionths of a Kelvin. This may be sufficient to seed the formation of large-scale structures in the universe, especially if there is a great deal of nonluminous or "dark matter" present in the universe. This invisible matter, whose nature is not yet confirmed, supposedly provided an added gravitational force needed to pull the gas together into galaxies within a reasonable time. We will explore the dark matter problem and ideas of structure formation in a later chapter.

Proponents of inflationary scenarios assert that the size of the observed fluctuations is consistent with their being produced by microscopic quantum effects that were magnified by the rapid pace of inflation or by gravitational waves generated during inflation itself. According to inflationary scenarios, the universe expanded rapidly for a brief instant just after the Big Bang. More detail will be introduced later.

7.6 Additional Evidence for the Big Bang

The cosmic microwave background radiation and its spectrum provide the strongest evidence in favor of the Big Bang theory. There are additional bits of evidence com-

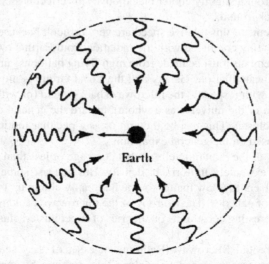

Fig. 7.11 Our motion through the microwave background.

ing from measurements on the abundance of helium and deuterium in the universe. Helium is formed from hydrogen in the interiors of stars during their lifetimes. On the other hand, the Big Bang cosmology predicts that helium was also formed from hydrogen in the early stages of the Big Bang. These circumstances led to a criterion for judging whether the Big Bang ever occurred. If nearly all the helium in the Universe were primordial, that would be good evidence in favor of the Big Bang. If it turned out that all the helium that exists today had been manufactured in the interiors of stars, none would be primordial, and confidence in the Big Bang theory would be weakened.

The answer to this problem was to measure the helium content of old and young stars. The young stars were formed from an interstellar medium containing primordial helium, if any, plus all the helium that was added to the universe subsequently in many generations of stellar evolution. The old stars were formed when the galaxy was young, before the interstellar medium had been enriched by helium formed in stellar interiors. Their helium content is primordial only. Therefore the comparison of the helium content in the two groups of stars tells how much helium is primordial, and how much has been added as the product of reactions in stellar interiors.

The helium content of young stars can be determined directly from the intensities of the helium absorption lines in their spectra. This absorption takes place only in the atmosphere of the star, so the intensity only gives the amount of helium in the star's outermost layer. However, in a young, unevolved star, the helium is dispersed uniformly; the amount in the atmosphere is an accurate indicator of the amount in the entire star. Only the hot stars—O and B type—can be used for this purpose, because helium lines appear only with significant intensity in the spectra of these stars. The spectroscopic studies indicate that about 30 percent of the mass of young stars consists of helium.

When we come to old stars, the helium absorption lines cannot be used in the same way to determine helium content, because a population of old stars does not include O and B types, which are massive and live only a short time–not more than 100 million years. But another method is available. The ages of old stars in globular clusters can be determined by fitting Hertzsprung-Russell (H-R) diagram computed for globular clusters of various ages to the observed H-R diagram. (H-R diagram is a diagram on which the absolute magnitude or luminosity of stars is plotted against spectral or surface temperature.) Computations on stellar structure show that the position of a star on the H-R diagram depends to some degree on its chemical composition. In particular, the luminosity (and so the star's position on the H-R diagram) is strongly dependent on the helium content of the star. By carefully matching the observed and theoretical H-R diagrams for a globular cluster, it is possible to determine not only the age of the stars in the cluster but also their helium content. Helium contents between 22% and 26% are found in this way. In other words, the helium content of old stars is a little less than the helium content of young stars, but close to it. This agrees with the prediction of the Big Bang theory that most of the helium in the universe was made shortly after the Big Bang; only a small amount was contributed subsequently by nuclear reactions in stars. The quantitative agreement between the predicted and observed amounts of primordial helium is impressive. These findings significantly strengthen the case for the Big Bang theory. The Big Bang nucleosynthesis of the light elements is important and will be explored in Chapter 11.

Astronomers have found direct evidence of primordial helium in the spectrum of ultraviolet light emitted by a distant quasar. As the emitted light traverses the vast expanse of space between the quasar and Earth, it encounters intergalactic helium and hydrogen. Gas completely ionized by the quasar light cannot absorb any more radiation, so the light passes unimpeded, as if it were traveling through a transparent medium. This appears to be the case for diffuse hydrogen that is easily stripped of its one electron.

It takes more energy to ionize a helium atom, which has two electrons. Although the quasar beacon fully ionizes most of the helium it encounters, some of the atoms manage to retain one of their electrons. When the radiation passes through singly ionized helium, the ions absorb light of a particular wavelength, leaving behind a fingerprint—a dark line, or gap, in the quasar's spectrum. But because of the redshift of light caused by the expansion of the universe, gaps due to helium ions at different distances along the line of sight to the quasar will appear at different wavelengths to an observer on Earth. Thus, the helium ions collectively create a series of dark absorption lines in the quasar spectrum.

The HUT (the Hopkins Ultraviolet Telescope), part of the Astro 2 Observatory that flew aboard the space shuttle in March 1994, recorded a series of such dark lines in the spectrum of the quasar HS 1700 + 64, which lies about 10 billion light-years from Earth. The singly ionized helium detected by HUT represents only a tiny fraction of the total amount of helium that resided in the early universe, because most of the gas is completely ionized.

The existence of deuterium provides even stronger support of the Big Bang. An ordinary hydrogen nucleus consists of a single proton. In deuterium, a proton and a neutron are bound together. Deuterium is a form of hydrogen and not some other element because the addition of a neutron to the nucleus does not alter its chemical properties. The nucleus still has a charge of +1, and it will still form an atom in which there is a single electron.

Deuterium is not very abundant in our universe. There is roughly about one deuterium atom for every 30,000 atoms of ordinary hydrogen. Yet the existence of even tiny quantities of deuterium provides scientists with significant evidence about the Big Bang. The deuterium nucleus is relatively fragile, and it cannot be created in stars. The high temperatures in stellar interiors would cause deuterium nuclei to break apart as soon as they were formed; thus, the only place that deuterium could have been created is in the Big Bang.

7.7 Problems

7.1. A galaxy has a recession speed of 13,000 km/s. What is its distance in mega-parsecs (Mpc)?

7.2. A galaxy at a distance of 300 Mpc has a recession speed of 21,000 km/s. If its recession speed has been constant over time, how long ago was the galaxy adjacent to our galaxy, the Milky Way?

7.3. Given the Hubble constant $H_0 = 65$ km/s per Mpc, calculate the Hubble time.

7.4. (a)Calculate the energy density u_r for the cosmic background radiation ($T = 2.7$ K), where $u_r = aT^4$ with $a = 4\sigma/c$ and σ is the Stefan-Boltzmann constant ($= 5.6697 \times 10^{-8}$ W/m$^2 \cdot$ K^4).

(b)Covert this energy density to an equivalent mass density ρ_r. This mass density is very useful in our study of the evolution (thermal history) of the universe.

7.5. Planck's law for the intensity of blackbody radiation is

$$I_\lambda = \frac{2hc^2/\lambda^5}{e^{hc/\lambda kT} - 1}.$$

As the universe expands with a scale factor ('radius') $R(t)$, the intensity varies as $I_\lambda \propto R^{-5}$ while the wavelength goes as $\lambda \propto R$.

(a) Show that $T \propto R^{-1}$ if the blackbody formula is to remain valid.
(b) At what wavelength does the blackbody curve reach a maximum for the observed 2.7 K background radiation?

References

Gulkis S, Lubin PM, Meyer SS, Silverberg RF January, 1990. The cosmic background explorer. Scientific American, January, 1990 (USA)

Harrison ER (1981) *Cosmology, the Science of the Universe* (Cambridge University Press UK)

Chapter 8
Big Bang Models

We now discuss the standard "gravity only" Big Bang models that are based on the Robertson-Walker metric and the Friedmann equations. After a brief review of "the cosmic fluid" and "fundamental observers," we give a concise derivation of the Robertson-Walker metric and examine some of its properties, then apply it to cosmic dynamics (the Friedmann equations). Finally we will discuss the recent discovery that the universe is accelerating, instead of slowing down (as expected from the standard "gravity only" model) and its implication for the evolution of the universe and other related problems.

8.1 The Cosmic Fluid and Fundamental Observers

Modern cosmology is based upon the description of the geometry of space-time given by general relativity. Thus we need first to write down the metric tensor of the universe. Obviously, this is almost an impossible task. Fortunately, the observational data indicate that the large-scale universe is both isotropic and homogeneous. The words "large-scale" refer to the scale of many superclusters of galaxies. Isotropic means that it looks the same in all directions, and homogeneous means that it would look the same from any vantage point. The best evidence for isotropy comes from measurements of cosmic microwave background radiation. Our first basic assumption is that the universe is isotropic and homogeneous. This assumption was dignified, in Chapter 7 as the cosmological principle. Thus, every point in the universe is equivalent to every other point; there are no preferred positions or directions in the cosmos.

The actual universe is clearly very nonhomogeneous on the small scale, and only rough limits can be put on its homogeneity on the large scale. Matter is concentrated in stars, which in turn are collected into galaxies or clusters of galaxies. We make no attempt to incorporate individual galaxies or clusters of galaxies into our description of the universe as a whole. Instead, we imagine matter in the universe as being smeared out into an idealized, smooth fluid, which is often called the "cosmic fluid," devoid of shear-viscous, bulk-viscous, and heat-conductive properties. This is an idealization and a good description so long as we take a large-scale view of the universe. In physics we encounter a similar situation when we study properties of a gas. To describe the properties of a gas we do not need to study the behavior of individual atoms and molecules. Instead we define various macroscopic quantities such as density, pressure, and temperature, and study the relations between them.

We call an observer who is at rest with respect to this cosmic fluid a fundamental observer. As the universe expands, the cosmic fluid shares in the expansion, and the fundamental observer will be co-moving with the fluid. Every co-moving observer in the cosmic fluid sees the same picture of the universe, i.e., in the co-moving frame of reference the universe looks isotropic and homogeneous. What is the form of the space-time metric in the co-moving frame? It is obvious that we cannot use the static Schwarzschild metric for an expanding universe that is certainly not static. Robertson and Walker found the form of the cosmological metric in the co-moving frame. This is known today as the Robertson-Walker metric. We just give its explicit form here and refer readers who are interested in its derivation to the book by Rindler (see References).

$$ds^2 = c^2 \, dt^2 - R^2(t) \left[\frac{dr^2}{1 - kr^2} + r^2 \, d\,\theta^2 + r^2 \sin^2 \theta \, d\varphi^2 \right] \qquad (8.1)$$

where t is the time in the co-moving frame, i.e., t is the proper time, and $R(t)$ is a dimensionless scale (or expansion) factor depending only on time t. At one instant of time t, the spatial metric is isotropic and homogeneous, which means that the three-dimensional space is isotropic and homogeneous. In other words, t is a cosmic standard time so that at every instant of t the universe looks isotropic and homogeneous.

Now let us take a close look at the line-element (1). The second term, in which k is a constant, measures distance in a spatial section of the space-time, which exists at an instant t of cosmic time. The physical distance in this space between two points separated by fixed coordinate intervals dr, $d\theta$, and $d\varphi$ varies with time in proportion to the scale factor $R(t)$, which depends only on time. As in the Schwarzschild line element, the coordinate r does not provide a linear measure of distance. However, t does measure a genuine time. The proper time τ measured by any observer whose spatial coordinates r, θ, and φ are fixed is clearly the same as t (for a particular galaxy, $dr = d\theta = d\varphi = 0$, then $ds = c \, dt$). Moreover, such an observer is moving through the space-time along a geodesic and is therefore in free fall, which would not be the case in Schwarzschild space-time. The sequence of spatial sections corresponding to successive instants of time can be thought of as a three-dimensional space that expands or contracts uniformly with time according to the variation of $R(t)$. The surfaces of constant r, θ, and φ expand or contract in the same way, like a grid of lines painted on the surface of an inflating balloon, and these coordinates are said to be co-moving.

It is important to keep in mind that there is a universal cosmic time t, which is the same for all observers at rest with respect to local matter. But in practice galaxies have random motions, so the local inertial frame has to be defined in terms of the average motion of galaxies over a sufficiently large region, or with respect to the microwave background radiation.

The parameter k is the curvature constant, which describes the geometry of space at a particular instant of time. We will see in the following section that $k > 0$ corresponds to a space of positive curvature, $k = 0$ to normal flat space, and $k < 0$ to a space of negative curvature. k can be taken as $+1$, 0, or -1 by a suitable rescaling of the radial coordinate r, which is a co-moving coordinate:

Why is the function $R(t)$ called a scale factor? If we determine the proper distance $c\,t$ between two galaxies from the equation $ds = 0$, we find

$$c\,dt = R(t)d\sigma$$

and so

$$c\Delta t = R(t)\Delta\sigma \tag{8.2}$$

where $d\sigma^2$ is the metric of the three-dimensional space at a fixed value of the cosmic time t,

$$d\sigma^2 = \frac{dr^2}{1 - kr^2} + r^2\left(d\theta^2 + \sin^2\theta\,d\varphi^2\right). \tag{8.3}$$

All measurements are understood to be made at the same epoch t. Now, since r is a co-moving label and remains fixed for each galaxy, and θ and φ remain fixed for isotropic motion, then $d\sigma$ is fixed, and the proper distance between two galaxies is just scaled by the function $R(t)$ as t varies. For this reason $R(t)$ is called the *scale-factor*; R increases or decreases as the universe expands or contracts.

8.2 Properties of the Robertson-Walker Metric

Now let us examine the nature of the geometry of the three-dimensional space at a fixed cosmic time t. From (8.1) the element of length dL is

$$dL = R^2(t)d\sigma^2 \equiv \gamma_{ij}dx^i\,dx^j \tag{8.4}$$

where $d\sigma^2$ is given by (8.3). The three-dimensional Riemann tensor corresponding to (8.4) is

$$R_{ijkl} = \frac{k}{R^2}(\gamma_{ik}\gamma_{jl} - \gamma_{jk}\gamma_{il}) \tag{8.5}$$

(calculated from (2.54) with indices restricted to run over 1, 2, 3), and the corresponding Ricci tensor and curvature scalar are

$$R_{ij} = \frac{2k}{R^2}\gamma_{ij} = \frac{1}{3}R_k^k\gamma_{ij}; \quad R_k^k = \frac{6k}{R^2}. \tag{8.6}$$

Thus the curvature properties of the space is specified by the constant $k/R^2(t)$. If $k > 0$ we have a space of constant positive curvature; if $k < 0$, it is a space of constant negative curvature. If $k = 0$ we have the usual flat space.

We first consider geometry of a space with k > 0 (i.e., having a constant positive curvature). We can gain additional intuitive insight into the geometry with the following new variables:

$$x = R\, r \sin\theta \cos\varphi; \quad y = R\, r \sin\theta \sin\varphi; \quad z = R\, r \cos\theta \tag{8.7}$$

We also introduce a redundant, extra variable x_4 by the relation

$$x^2 + y^2 + z^2 + x_4^2 \equiv R^2/k \tag{8.8}$$

from which we have

$$R^2\, r^2 + x_4^2 = R^2/k. \tag{8.9}$$

Since R^2/k is a constant (at one instant of time t) it follows that

$$R^2\, r\, dr + x_4 dx_4 = 0$$

or

$$\left(dx_4\right)^2 = \frac{k\, r^2\, r^2\, dr^2}{1 - kr^2}. \tag{8.10}$$

Using this result we can now rewrite dL^2 (8.4), as

$$dL^2 = R^2 \left[dr^2 + r^2\, d\theta^2 + r^2 \sin^2\theta\, d\varphi^2 \right] + \frac{kR^2\, r^2\, dr^2}{1 - kr^2} \tag{8.11}$$
$$= dx^2 + dy^2 + dz^2 + dx_4^2$$

The right-hand side of the above equation is the expression for the line element of a four-dimensional space with Cartesian coordinates x, y, z, and x_4, and (8.8) implies that the variables are confined to the surface of a hypersphere with radius R/k. Thus, for $k > 0$, the universe is a three-space embedded in a (fictitious) four-dimensional Euclidean space. The circumference of a circle in the "spherical" coordinates r, θ, φ is equal to $2\pi r$, and the surface of a sphere to $4\pi r^2$. But the radius of a circle (or sphere) is equal to

$$\int_0^r \frac{dr}{\sqrt{1 - kr^2}} = \frac{1}{\sqrt{k}} \sin^{-1}\left(\sqrt{k}r\right) \tag{8.12}$$

i.e., is larger than r. Thus the ratio of circumference to radius in this space is less than 2π. The volume of this space is finite. To show this, let us introduce in place of the coordinate r the "angle" variable

$$\sin\chi = \sqrt{k}\, r. \tag{8.13}$$

The range of χ goes between the limits 0 and χ, which is found from the restriction $0 \leq r \leq 1$. In terms of this new variable, dL^2 takes the new form

$$dL^2 = a^2 \left[d\chi^2 + \sin^2\chi\, (d\theta^2 + \sin^2\theta\, d\varphi^2) \right] \tag{8.14}$$

where $a^2 = R^2/k$. The coordinate χ determines the distance from the origin, given by $a\chi$, and the total volume of space is equal to

$$V = \int_0^{2\pi} \int_0^\pi \int_0^\pi a^3 \sin^2 \chi \sin \theta \, d\chi \, d\theta d\varphi = 2\pi^2 \, a^3 = 2\pi^2 \, k^{-3/2} r^3. \quad (8.15)$$

Thus, a space of positive curvature turns out to be closed on itself. Its volume is finite, though it has no boundaries.

For a space with $k < 0$ (i.e., having a constant negative curvature), the element of length has, in coordinates r, θ, and φ, the form

$$dL^2 = R^2(t) \left[\frac{dr^2}{1 + |k| r^2} + r^2 \left(d\theta^2 + \sin^2\theta \, d\varphi^2 \right) \right] \quad (8.16)$$

where the coordinate r can go through all values from 0 to ∞. The ratio of the circumference of a circle to its radius is now greater than 2π. Corresponding to (8.13) for the angle variable χ, we have now

$$\sinh \chi = |k|^{1/2} r; \quad 0 \le \chi < \infty. \quad (8.17)$$

Then,

$$dL^2 = \left(R^2/|k| \right) \left[d\chi^2 + \sinh^2 \chi \left(d\theta^2 + \sin^2 \theta d\varphi^2 \right) \right] \quad (8.18)$$

and the surface of a sphere is now equal to

$$4\pi^2 r^2 = 4\pi^2 \left(R^2/|k| \right) \sinh \chi. \quad (8.19)$$

We see that as we move away from the origin (increasing χ), the space increases without limit. The volume of this space is clearly infinite. Thus a universe of negative curvature is an open universe.

For a space with $k = 0$, the element of length reduces to

$$dL^2 = R^2(t) \left[dr^2 + r^2 \left(d\theta^2 + \sin^2\theta d\varphi^2 \right) \right]. \quad (8.20)$$

The space is, clearly, Euclidean; the volume is again infinite.

Finally, we emphasize that the Robertson-Walker metric is a consequence only of the symmetry of three-dimensional position space, expressed in a four-dimensional language. The number of unknowns in the metric tensor is reduced from 10 to the single function, function $R(t)$ and the discrete parameter k.

Now let us see how simply the function $R(t)$ describes the expansion of the universe. Specifically we will discuss some of the important observational features of a typical Robertson-Walker space-time. These features show how a non-Euclidean geometry can differ substantially with conclusions based on naive Euclidean concepts.

The way in which the scale-factor $R(t)$ varies with time has to be determined by substituting the Robertson-Walker metric into Einstein's field equations; and we shall learn there that $R(t)$ satisfies a set of equations known as Friedmann's equations. A variety of model universes can be introduced, characterized by different values of the scale-factor $R(t)$.

Before turning our attention to cosmic dynamics, let us see how Hubble's law is accounted for by the Robertson-Walker metric. Assume that our galaxy and those we observe are co-moving, so that their spatial coordinates are fixed. Then the physical distance (often called the proper distance) between two galaxies separated by a coordinate distance d_0 is $d(t) = R(t)d_0$ at a given cosmic time t. And their relative velocity is

$$v = \frac{d}{dt}d(t) = \frac{\dot{R}(t)}{R(t)}d(t) = H(t)d(t).$$

It says that at any given cosmic time t the speed of a galaxy relative to us is proportional to its distance from us. This is simply Hubble's law, with the Hubble constant given by

$$H(t) = \dot{R}(t)/R(t). \tag{8.21}$$

The recession speed can be measured as the redshift of spectral lines. Redshift is also a measure of the scale factor. To see this, we consider the case of a wave emitted by a co-moving galaxy, say at $r = r_e$ and $\theta = \varphi = 0$, and received by a co-moving observer (us) at $r = 0$. The light ray moves along a null geodesic whose equation, according to (8.1), is

$$0 = c^2 dt^2 - R^2(t)dr^2 / \left(1 - kr^2\right)$$

from which we get

$$cdt = -R(t)dr / \left(1 - kr^2\right)^{1/2},$$

with the "−" sign corresponding to a ray moving toward the origin (r decreases as t increases). If a wave crest is emitted at time t_e and received at t_0, then

$$\int_{t_e}^{t_0} \frac{dt}{R(t)} = -\frac{1}{c}\int_0^{r_e} \frac{dr}{\sqrt{1 - kr^2}} = d_0$$

where d_0 is independent of both t_e and t_0. If the following crest is emitted at time $t_e + \Delta t_e$, then

$$\int_{t_e + \Delta t_e}^{t_0 + \Delta t_0} \frac{dt}{R(t)} = d_0 + \frac{\Delta t_0}{R(t_0)} - \frac{\Delta t_e}{R(t_e)} = d_0$$

and so the observed frequency and wavelength are related to those of the emitted wave by

$$\frac{v_0}{v_e} = \frac{R(t_e)}{R(t_0)} \quad \text{or} \quad \frac{\lambda_0}{\lambda_e} = \frac{R(t_0)}{R(t_e)}.$$

As seen by a co-moving observer, therefore, the wavelength of a photon changes in proportion to the scale factor. Now the redshift z can be written in terms of the scale factor

$$z = \frac{\lambda_0 - \lambda_e}{\lambda_e} = \frac{R(t_0)}{R(t_e)} - 1 \tag{8.22}$$

In an expanding universe, $R(t_0) > R(t_e)$, so that z is positive, as expected.

It is worth emphasizing that our derivation above shows that the redshift effect arises from the passage of light through non-Euclidean space-time. It does not arise from the Doppler effect. We referred to this as cosmological redshift in Chapter 7.

If the scale factor at the present epoch $R(t_0)$ is set equal to unity, then

$$z = \frac{1}{R(t_e)} - 1 \text{ or } R(t_e) = \frac{1}{1+z} \tag{8.23}$$

Red shift is simply a measure of the scale factor of the universe at t_e (i.e., when the source emitted its radiation). For example, when we observe a galaxy with $z = 1$, the scale factor of the universe at t_e was $R(t_e) = 0.5$, i.e., the universe was half its present size. But we have no information about when the light was emitted. If we did, we could determine $R(t_e)$ directly from observation. We have to have some theory of cosmic dynamics in order to determine $R(t)$. This is the subject of the next section.

If the observed cosmological redshift is small, so that t_e is (cosmologically speaking) not much earlier than t_0, then we can expand $R(t_e)$ about t_0

$$R(t_e) = R(t_0) + (t_e - t_0)\dot{R}(t_0) + (1/2)(t_e - t_0)^2 \ddot{R}(t_0) + \ldots$$
$$= R(t_0)\left[1 + H_0(t_e - t_0) - (1/2)q_0 H_0^2(t_e - t_0)^2 + \ldots\right]$$

where

$$H(t_0) = \dot{R}(t_0)/R(t_0), \quad \text{and} \quad q_0 = -\ddot{R}(t_0)R(t_0)/\dot{R}^2(t_0) = -\ddot{R}(t_0)/[R(t_0)H_0^2],$$

the dimensionless quantity q_0 is called the deceleration parameter. The redshift z can now be expanded in powers of $t_0 - t_e$:

$$z = H_0(t_0 - t_e) + (1 + q_0/2)H_o^2(t_0 - t_e)^2 + \ldots$$

and, conversely, the time of light travel $t_0 - t_e$ may be expanded as a function of z

$$t_0 - t_e = \frac{1}{H_0}\left[z - \left(1 + q_0/2\right)z^2 + \ldots\right].$$

These formulae are very useful, but it should not be forgotten that they are only valid for small z.

8.3 Cosmic Dynamics and Friedmann's Equations

The uniform model universes based on the Robertson-Walker line element are characterized by their scale factors $R(t)$ and the curvature index k. But we have so far been concerned with cosmic kinematics that do not tell us how the scale factor $R(t)$ varies with time t; thus we do not know the rate at which the universe expands as given by the scale factor $R(t)$. We also do not know whether the universe is open or closed as indicated by the curvature parameter k. To find answers to these questions we need a dynamic theory, i.e., to combine the isotropic, homogeneous Robertson-Walker line element with Einstein's field equations. This procedure will give us the dynamical equations satisfied by the scale factor $R(t)$.

We first compute the Einstein tensor. If we label our coordinates accordingly as $x^o = ct, x^1 = r, x^2 = \theta, x^3 = \varphi$, then the nonzero components of $g_{\mu\nu}$ and $g^{\mu\nu}$ are:

$$g_{oo} = 1 = g^{oo}, g_{11} = -\frac{R^2}{1 - kr^2} = \left(g^{11}\right)^{-1}, g_{22} = -R^2 r^2 = \left(g^{22}\right)^{-1}$$

$$g_{33} = -R^2 r^2 \sin^2\theta = \left(g^{33}\right)^{-1}, \sqrt{-g} = \frac{R^3 r^2 \sin\theta}{\sqrt{1 - kr^2}}. \tag{8.24}$$

The nonzero components of $\Gamma^\mu_{\nu\alpha}$ are then as follows:

$$\Gamma^1_{01} = \Gamma^2_{02} = \Gamma^3_{03} = \frac{1}{c}\frac{\dot{R}}{R}, \left(\Gamma^i_{0j} = -\frac{1}{2}\frac{\partial}{\partial t}g_{ij}\right)$$

$$\Gamma^0_{11} = \frac{R\dot{R}}{c(1 - kr^2)}, \quad \Gamma^0_{22} = \frac{r^2 R\dot{R}}{c}, \quad \Gamma^0_{33} = \frac{S\dot{R}r^2\sin^2\theta}{c}; \left(\Gamma^0_{ij} = -\frac{1}{2}\frac{\partial}{\partial t}g_{ij}\right)$$

$$\Gamma^1_{11} = \frac{kr}{1 - kr^2}, \quad \Gamma^1_{22} = -r\left(1 - kr^2\right), \quad \Gamma^1_{33} = -r\left(1 - kr^2\right)\sin^2\theta$$

$$\Gamma^2_{12} = \Gamma^3_{13} = \frac{1}{r}, \quad \Gamma^2_{33} = -\sin\theta\cos\theta, \quad \Gamma^3_{23} = \cot\theta. \tag{8.25}$$

A dot, as before, denotes differentiation with respect to time. We now calculate the Ricci tensor, which may be put in the following form:

$$R_{\mu\nu} = \frac{\partial^2 \ln\sqrt{-g}}{\partial x^\mu \partial x^\nu} - \frac{\partial \Gamma^\lambda_{\mu\nu}}{\partial x^\lambda} + \Gamma^\beta_{\mu\alpha}\Gamma^\alpha_{\nu\beta} - \Gamma^\lambda_{\mu\nu}\frac{\partial \ln\sqrt{-g}}{\partial x^\lambda}$$

Straightforward but tedious calculation gives the following nonzero components of the Ricci tensor:

$$R^o_o = \frac{3}{c^2}\frac{\ddot{R}}{R}, R^1_1 = R^2_2 = R^3_3 = \frac{1}{c^2}\left(\frac{\ddot{R}}{R} + \frac{2\dot{R}^2 + 2kc^2}{R^2}\right). \tag{8.26}$$

From these we get the scalar curvature S

$$S = R^k_k = \frac{6}{c^2}\left(\frac{\ddot{R}}{R} + \frac{\dot{R}^2 + kc^2}{R^2}\right) \tag{8.27}$$

and hence the Einstein tensor $G^\mu_{\ \mu}$:

$$G^1_1 = R^1_1 - \frac{1}{2}S = -\frac{1}{c^2}\left(2\frac{\ddot{R}}{R} + \frac{\dot{R}^2 + kc^2}{R^2}\right) = G^2_2 = G^3_3 \tag{8.28}$$

$$G^0_0 = R^0_0 - \frac{1}{2}S = -\frac{3}{c^2}\left(\frac{\dot{R}^2 + kc^2}{R^2}\right). \tag{8.29}$$

We now apply the Einstein equations:

$$R_\mu^\nu - \frac{1}{2}\delta_\mu^\nu R_\mu^\nu = \frac{8\pi G}{c^4}T_\mu^\nu. \tag{8.30}$$

The "time-time" component (i.e., $\mu = 0, \nu = 0$) gives us

$$2\frac{\ddot{R}}{R} + \frac{\dot{R}^2 + kc^2}{R^2} = \frac{8\pi G}{c^2}T_1^1 = \frac{8\pi G}{c^2}T_2^2 = \frac{8\pi G}{c^2}T_3^3 \tag{8.31}$$

while the "space-space" component gives us

$$\frac{\dot{R}^2 + kc^2}{R^2} = \frac{8\pi G}{3c^2}T_o^o. \tag{8.32}$$

Note that the three nontrivial spatial equations are equivalent. This is essentially due to the homogeneity and isotropy of the Robertson-Walker line element.

We now need the energy tensor $T^\mu_{\ \mu}$ to describe the cosmic fluid. As we idealize the universe, and model it by a simple macroscopic ideal fluid devoid of viscous and heat-conductive properties, its energy tensor is then that of a perfect fluid, so

$$T_\mu^\nu = -p\delta_\mu^\nu + (p + \rho)U^\nu U_\mu; \quad U_\mu U^\mu = 1 \tag{8.33}$$

where p is the pressure, ρ is the energy-density of the cosmic fluid, and U_μ is the (covariant) world velocity of the fluid particles (galaxies). In the co-moving frame that we have chosen, the fluid is at rest, so

$$U^i = (1, 0, 0, 0) \tag{8.34}$$

and

$$T_o^o = \rho c^2, \quad T_1^1 = T_2^2 = T_3^3 = -p. \tag{8.35}$$

Substituting these into (8.31) and (8.32) we obtain

$$2\frac{\ddot{R}}{R} + \frac{\dot{R}^2 + kc^2}{R^2} = -\frac{8\pi G}{c^2}p \tag{8.36}$$

and

$$\frac{\dot{R}^2 + kc^2}{R^2} = \frac{8\pi G}{3}\rho. \tag{8.37}$$

Finally, we have the equations of motion of the fluid particles

$$T_{\lambda;\nu}^\nu = 0. \tag{8.38}$$

Writing this out in full, we have

$$\frac{\partial p}{\partial x^\lambda} - \frac{\partial}{\partial x^\nu}\left[(p + \rho)U^\nu U_\lambda\right] - \Gamma_{\sigma\nu}^\nu(p + \rho)U^\sigma U_\lambda + \Gamma_{\nu\lambda}^\sigma(p + \rho)U^\nu U_\sigma = 0.$$

With the aid of (8.25) and (8.26) the above equation reduces to

$$\frac{\partial p}{\partial x^\lambda} - \frac{d}{dt}\left[(p+\rho)U_\lambda\right] - 3(p+\rho)U_\lambda\frac{\dot{R}}{R} = 0.$$

The components $\lambda = 1, 2, 3$ of the above equation give the trivial equation $0 = 0$; the component $\lambda = 0$ gives the nontrivial equation

$$\dot{\rho} + 3(p+\rho)\frac{\dot{R}}{R} = 0. \tag{8.39}$$

The three equations (8.36), (8.37), and (8.39) are not all independent; for instance, taking a derivative of (8.37) and using (8.39) we can produce (8.36). Thus, it is sufficient to retain any two of these three equations. We shall keep (8.36) and (8.37) and refer to them as the Friedmann equations, after A. Friedmann, who first obtained these equations.

Now, in our smooth fluid approximation a velocity field such as (8.34) represents an orderly motion with no pressure. That is, we have in this case the system of galaxies behaving like (incoherent) dust. With this approximation, (8.39) becomes

$$\frac{d}{dt}\left(\rho R^3\right) = 0,$$

which integrates to

$$\rho R^3 = constant = \rho_0 R_0^3, \tag{8.40}$$

ρ_0 and R_0 being the values of ρ and R at the present epoch. This is the so-called Friedmann integral, which shows that matter density varies in time as R^{-3}, and the quantity of matter contained in a co-moving volume-element is constant during the expansion. This discussion is relevant to describe a matter-dominated epoch. For a radiation-dominated epoch, pressure cannot be ignored and is related to density by $p = c^2/3$.

In the present chapter we will consider the matter-dominated epochs, and so the two Friedmann equations become:

$$2\frac{\ddot{R}}{R} + \frac{\dot{R}^2 + kc^2}{R^2} = 0 \tag{8.41}$$

$$\frac{\dot{R}^2 + kc^2}{R^2} = \frac{8\pi G\rho_o}{3}\frac{R_o^3}{R^3}. \tag{8.42}$$

8.4 The Solutions of Friedmann's Equations

We now consider the solutions of the Friedmann equations for the three cases $k = 0$, 1, and -1. Before we consider the three cases separately, let us do some preparation work. First, recall that the Hubble constant $H(t)$ is defined by

$$H(t) \equiv \dot{R}(t)/R(t),$$

and we denote its present-day value by $H_o = H(t_o)$. Now, applying (8.42) to the present epoch and rewriting it in terms of H_o, we get

$$\frac{k}{R_o^2} = \frac{8\pi G \rho_o}{3c^2} - \frac{H_o^2}{c^2} = \frac{8\pi G}{3c^2}\left(\rho_o - \frac{3H_o^2}{8\pi G}\right). \tag{8.43}$$

Hence $k > 0$, $k = 0$ or $k < 0$ as $\rho_o > \rho_c$, $\rho_o = \rho_c$, or $\rho_o < \rho_c$ respectively, where ρ_c is called the critical density given by

$$\rho_c = 3H_o^{\,2}/8\pi G. \tag{8.44}$$

With the range of values of H_o known today, we have $\rho_c = 2 \times 10^{-29} h_o^2\,\text{g cm}^{-3}$, where h_o lies in the range $0.5 < h_o < 1$.

Similarly, the present-day value q_0 of the deceleration parameter $q(t)$ can also be expressed in terms of H_0 and ρ_c

$$q_o = \frac{4\pi G \rho_o}{3H_o^{\,2}} = \frac{\rho_o}{2\rho_c}. \tag{8.45}$$

8.4.1 Flat Model ($k = 0$)

For $k = 0$, (8.43) gives $\rho_o = \rho_c$; then (8.45) gives $q_o = 1/2$. Now let us return to the Friedmann equation (8.63), which becomes

$$\dot{R}^2 = \frac{8\pi G \rho_o}{3}\frac{R_o^3}{R} = \frac{A^2}{R}; \quad A^2 = \frac{8\pi G \rho_o R_o^3}{3}. \tag{8.46}$$

Integration gives

$$R(t) = (3A/2)^{2/3}t^{2/3}. \tag{8.47}$$

The integration constant has been set equal to zero by assuming that $R = 0$ at $t = 0$. We also get

$$t_o = \frac{2}{3H_o}. \tag{8.48}$$

Figure 8.1 illustrates this solution. Point A on the t-axis denotes the present epoch. The ordinates at A gives the present value of the scale factor, $PA = R_o$. The present value of the Hubble constant H_o is given by the ratio $1/AB$, where B is the intersection point of the tangent to the $R(t)$ curve at P with the t-axis. The age of the universe, represented by $0A$, is 2/3 of the intercept AB. Note that $\dot{R} \to 0$ as $t \to \infty$.

This model is also known as the Einstein-de Sitter model—Einstein and de Sitter gave it in a joint paper in 1932.

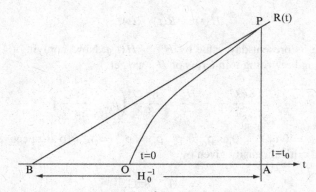

Fig. 8.1 A schematic graph of R(t) as a function of t for the flat model.

8.4.2 Closed Model (k = 1)

For k = 1, (8.44) gives $\rho_o > \rho_c$, then (8.45) gives $q_o > 1/2$. Now (8.41) and (8.42) become in this case

$$2\frac{\ddot{R}}{R} + \frac{\dot{R}^2 + c^2}{R^2} = 0 \tag{8.49}$$

$$\frac{\dot{R}^2 + c^2}{R^2} = \frac{A^2}{R^3} \tag{8.50}$$

where $A^2 = 8\pi G_o R_o^3/3$. Now (8.50) gives

$$\frac{dR}{dt} = c\sqrt{\frac{B^2 - R}{R}}; \quad B^2 = \frac{A^2}{c^2}$$

from which we have

$$ct = \int_0^R \sqrt{\frac{R}{B^2 - R}} dR.$$

Make the substitution

$$R = B^2 \sin^2(\phi/2) = \frac{1}{2}B^2(1 - \cos\phi).$$

The integral then becomes

$$ct = \frac{1}{2}B^2 \int (1 - \cos\phi)d\phi = \frac{1}{2}B^2(\phi - \sin\phi)$$

where we have taken R = 0 at t = 0 ($\varphi = 0$). Now we have

$$R = \frac{1}{2}B^2(1 - \cos\phi), \quad ct = \frac{1}{2}B^2(\phi - \sin\phi) \tag{8.51}$$

and these two equations give $R(t)$ via the parameter φ. The graph of $R(t)$ is a cycloid. In closed models, therefore, expansion is followed by contraction and R decreases to zero. The value $R = 0$ is reached when $\varphi = 2\pi$; that is, when $t = t_L = B^2\pi/c$. Now B^2 can be expressed in terms of q_o and H_o as (to be shown later)

$$B^2 = \frac{2q_o}{(2q_o - 1)^{3/2}} \frac{c}{H_o}$$ (8.52)

so

$$t = t_L = \frac{2\pi q_o}{(2q_o - 1)^{3/2}} \frac{1}{H_o}.$$ (8.53)

The quantity t_L may be termed the "lifespan" of this universe. For $q_o = 1, t_L = 2\pi/H_o$.

To derive (8.52), we apply (8.41) and (8.42) to the present epoch and express them in terms of q_o and H_o as

$$c/R_0^2 = (2q_o - 1) H_o^2$$ (8.54)

$$\rho_o = \left(H_o^2 + \frac{c^2}{R_o^2}\right) \frac{3}{8\pi G} = \frac{3H_o^2}{4\pi G} q_o.$$

Substituting R_o and ρ_o from these two equations into B^2

$$B^2 = \frac{A^2}{c^2} = \frac{8\pi G\rho_o R_o^3}{3c^2},$$

we obtain, after some straightforward manipulations, the desired result, namely (8.52).

Applying (8.51) to the present epoch we get

$$R_o = \frac{1}{2} B^2 (1 - \cos\varphi_0).$$ (8.55)

Substituting B^2 from (8.52) and R_o from (8.54) into the above equation we get

$$\frac{1}{2} \frac{2q_o}{(2q_o - 1)^{3/2}} \frac{c}{H_o} (1 - \cos\varphi_0) = \frac{c}{H_o} \frac{1}{(2q_o - 1)^{3/2}}.$$

From this we obtain, after simplification,

$$\cos\phi_o = \frac{1 - q_o}{q_o}; \quad \sin\varphi_0 = \frac{\sqrt{2q_o - 1}}{q_o}.$$

We therefore get from (8.51)

$$t_o = \frac{B^2}{2c} (\varphi_0 - \sin\varphi_0) = \frac{q_o}{(2q_o - 1)^{3/2}} \left[\cos^{-1}\left(\frac{1 - q_o}{q_o}\right) - \frac{\sqrt{2q_o - 1}}{q_o}\right] \frac{1}{H_o}.$$ (8.56)

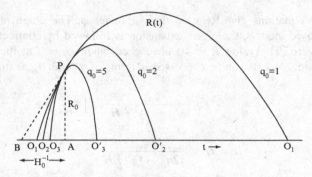

Fig. 8.2 Closed model. The schematic R(t) curves for $q_o = 1, 2, 5$.

For example, for $q_o = 1$, we get

$$t_o = \left(\frac{\pi}{2} - 1\right) H_o^{-1}. \tag{8.57}$$

From (8.51) we see that R reaches a maximum value at $\varphi = \pi$:

$$R = R_{\max} = B^2 = \frac{2q_o}{(2q_o - 1)^{3/2}} \frac{c}{H_o}. \tag{8.58}$$

Thus, for $q_o = 1$, the universe expands to twice its present size. Figure 8.2 illustrates the function $R(t)$ for the closed models for a number of parameter values q_o. All curves have been scaled to touch at P, the present point, and they all have the common tangent PB. The intercept $AB = H_o^{-1}$. Note that as q_o increases, the curves for $R(t)$ intercept the past section of the t-line at points $0_1, 0_2, \ldots$ lying closer and closer to A, that is, the age of the universe is reduced if q_o is increased.

8.4.3 Open Model ($k = -1$)

For $k = -1$, (8.43) and (8.45) give $\rho_o > \rho_c$, $q_o < 1/2$; and (8.41) and (8.42) become in this case

$$2\frac{\ddot{R}}{R} + \frac{\dot{R}^2 - c^2}{R^2} = 0 \tag{8.59}$$

$$\frac{\dot{R}^2 - c^2}{R^2} - \frac{8\pi G \rho_o R_o^3}{3R^3} = 0. \tag{8.60}$$

Applying them to the present epoch, we can rewrite (8.59) and (8.60) in terms of q_o and H_o as

$$\frac{c^2}{R_o^2} = (1 - 2q_o)H_o^2; \quad \rho_o = \frac{3H_o^2}{4\pi G} q_o. \tag{8.61}$$

The Equation (8.60) gives

$$\dot{R}^2 = \frac{8\pi G\rho_o R_o^3}{3}\frac{1}{R} + c^2 = c^2\left(1 + \frac{\alpha}{R}\right) \tag{8.62}$$

$$\alpha = \frac{8\pi G\rho_o R_o^3}{3c^2} = \frac{2q_o}{(1-2q_o)^{3/2}}\frac{c}{H_o}. \tag{8.63}$$

From (8.61) we get

$$dR/dt = c(1+\alpha/R)^{1/2},$$

so

$$ct = \int_o^R \sqrt{\frac{R}{\alpha + R}}\,dR.$$

Make the substitution

$$R = \alpha\sinh^2(\varphi/2) = \frac{1}{2}\alpha(\cosh\varphi - 1) \tag{8.64}$$

Then the integral becomes

$$ct = \frac{1}{2}\int(\cosh\varphi - 1)d\varphi = \frac{1}{2}(\sinh\varphi - \varphi).$$

Again, as in the previous two cases, we have taken R = 0 at $t = 0(\varphi = 0)$. Thus we have

$$R(t) = \frac{1}{2}\alpha(\cosh\varphi - 1), \quad ct = \frac{1}{2}\alpha(\sinh\varphi - \varphi). \tag{8.65}$$

These give $R(t)$ via the parameter φ. Since $R = \sinh\varphi/(\cosh\varphi - 1)$ so $R \to 1$ as φ (and hence t) $\to \infty$. Like the Einstein-de Sitter model, this model continues to expand forever.

Similar to the case of k = 1, the present value of φ is given by

$$\cosh\varphi_0 = \frac{1-q_o}{q_o}, \quad \sinh\varphi_o = \frac{\sqrt{1-2q_o}}{q_o}. \tag{8.66}$$

The present value of t is given by

$$t_o = \frac{\alpha}{2c}(\sinh\varphi_0 - \varphi_0) = \frac{q_o}{(1-q_o)^{3/2}}\left[\frac{\sqrt{1-2q_o}}{q_o} - \ln\left(\frac{1-q_o+\sqrt{1-2q_o}}{Q_o}\right)\right]H_o^{-1}. \tag{8.67}$$

The behavior of $R(t)$ is illustrated in Fig. 8.3. As in Fig. 8.2, all curves have the same value of H_o at P. The age of the universe is seen to increase as q_o decreases, being maximum ($= H_o^{-1}$) for $q_o = 0$.

We plot the three possible evolutions of the universe in Fig. 8.4. This graph also shows the newly discovered accelerating universe, and we will discuss this recent discovery and its implication for the evolution of the universe and other related problems in section 8.7.

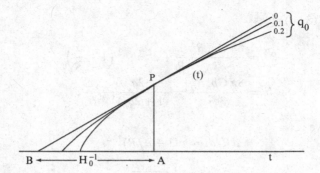

Fig. 8.3 Open model. The schematic R(t) curves for $q_0 = 0$, 0.1, and 0.2.

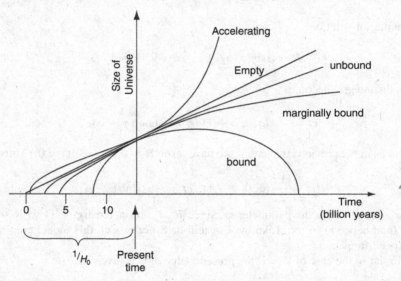

Fig. 8.4 The three possible evolutions of the universe and the newly discovered accelerating universe. (a) An unstable equilibrium. (b) A metastable equilibrium.

8.5 Dark Matter and the Fate of the Universe

According to the standard "gravity only" model, the average density of matter in the universe will determine its future. If the average density of matter is less than the critical density ρ_c, the expansion of the universe will continue forever, and we say that the universe is open or unbounded. Conversely, if the average density is greater than ρ_c, the gravity will be strong enough to eventually halt the expansion of the universe. At some point, the universe will reach a maximum state and then begin contracting. In such a case, we say that the universe in bounded, or closed. At present the observed mass density of the universe is much less than the critical density ρ_c. This discrepancy leads to the so-called "dark matter problem," originally known as "the problem of missing mass."

The Swiss astronomer Zwicky investigated the missing mass problem for the first time in the 1930s. He measured the mass of clusters of galaxies in two ways. Because there is a definite relation between the luminosity of a galaxy and its mass, Zwicky determined the mass of galaxies from their luminosity measurements. Then, by adding up all the masses of the member galaxies, he obtained the total mass of the cluster. Mass of the cluster can also be determined by measuring the relative velocities among the galaxies, since the mean relative velocity is determined by the mass of the cluster. Zwicky discovered that the masses determined by these two methods differed greatly. For example, the dynamical mass of the Coma cluster is 400 times the luminosity mass! Zwicky believed that there must be a large amount of invisible matter within the cluster. Nothing was known at that time about what sort of matter was contributing to this invisible mass, so it was called "missing mass."

Zwicky's bold conjecture was not well received. Since the 1950s, observational evidence supporting Zwicky's "missing mass" conjecture began to increase. The first decisive evidence came from the rotation curves of galaxies, which showed the velocity of matter rotating in a spiral disk as a function of the radius from the center. The individual stars in orbit obey Kepler's third law; if $M(r)$ is the mass of a galaxy within a radius r, then we have

$$\frac{v^2}{r} = \frac{GM(r)}{r^2} \quad \text{or} \quad v = \sqrt{\frac{GM(r)}{r}},$$

which simply says that the centripetal force of a orbiting star is provided by gravitational force. Thus, the farther away a star is from the center, the smaller is its rotational velocity (Fig. 8.5). However, the observed rotational curve of a galaxy is completely different. Figure 8.6 is the measured rotation curves of galaxies; the rotating velocity of bodies is independent of their distance. The only possible interpretation of this result is that the space surrounding the galaxy is not empty; rather, it is a halo with a considerable mass. It does not emit light, so it is invisible. Today, it is generally accepted that up to 90% of the universe is made of invisible "dark matter."

What, then, is dark matter? Diffuse gas was the first candidate for consideration. Our galaxy contains many gas clouds. Could there be gaseous matter in large amounts in intergalactic space? Many 21-cm observations show that the density of intergalactic hydrogen gas is less than 10^{-2} atoms per cubic centimeter–not enough

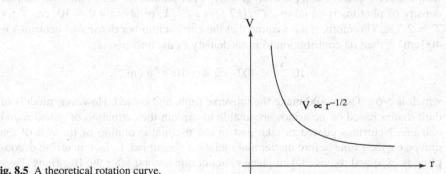

Fig. 8.5 A theoretical rotation curve.

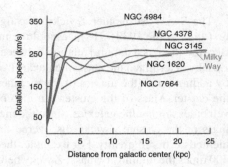

Fig. 8.6 Rotation curve of galaxies.

to meet the deficit found by Zwicky. Indications from optical methods show that the density of intergalactic hydrogen cannot exceed 10^{-12} atoms per cm^3. The optical methods can also detect other intergalactic atoms; the result has been entirely negative.

Yet perhaps the intergalactic gas is in the ionized form. A high-temperature ionized gas will emit X-rays, and X-rays have indeed been found in clusters of galaxies. But the density found is far from enough to account for the deficit mass.

Recent experiments indicate that neutrinos may have a rest mass of $\sim 6 \times 10^{-32}$ g. If this is accurate, then neutrinos may just be the missing matter, because there are enormous numbers of neutrinos in the present universe. We can make a simple calculation for that number. We shall see in the next chapter that the annihilation of electrons and positrons in the early universe produced a sea of neutrinos and antineutrinos. At that time their numbers would have been comparable to the number of "cosmic" photons. In any volume R^3 that follows the expansion of the universe, the numbers of both photons and neutrinos would have been conserved (assuming negligible rates of net annihilations); so the number density of neutrinos and antineutrinos today must still be comparable to the number density of photons. Hence we can calculate the latter value and thereby get an estimate of the number density of neutrinos and antineutrinos.

The cosmic photons are from the cosmic background radiation that is a relic from the Big Bang. The cosmic background radiation has a blackbody spectrum at temperature $T = 2.7$ K. Now, the energy density associated with blackbody radiation of temperature T is aT^4, and the average energy per photon is $\sim kT$. Thus, the number density of photons is equal to $aT^4/(kT) = aT^3/k$, or about $4.0 \times 10^2 \, cm^{-3}$ for $T = 2.7$ K. Therefore, if we assume that the current number density of neutrinos is $400 \, cm^{-3}$, then its contribution to mass density of the universe is

$$6 \times 10^{-32} \times 400 = 2.4 \times 10^{-29} \, g \, cm^{-3}$$

which is $> \rho_c$. This could make the universe finite and closed. However, models of dark matter based on neutrinos are unable to explain the formation of galaxies and clusters. Neutrinos ceased to take part in the thermal evolution of the rest of the universe a long time before matter and radiation decoupled. In fact, neutrino decoupling is expected to have taken place at about one second after the Big Bang. Ever since that time, they have hardly reacted with matter at all. Their very high speed causes them to resist clumping together, and their inherent gravity has a smoothing effect on the distribution of matter, pulling apart the density fluctuations that occur in the baryonic matter. If the neutrinos have zero mass, then they provide very little mass-energy density, since (because of the expansion of the universe) their kinetic energies are tiny today.

It is still an open question as to whether neutrinos have a rest mass or not. But discussion of neutrinos' rest mass has changed the perspective on the dark matter problem. Now it is generally regarded as an area to be opened up by the combined efforts of astrophysics and particle physics. Both theories of supergravity and supersymmetry predict the existence of many new particles, and these new particles have very weak interaction that cannot be detected in today's laboratories. Some physicists have speculated already that the dark matter in the universe may consist of just these particles. We will revisit the dark matter problem in Chapter 11.

Instead of trying to find all of the matter needed to close the universe, we can look for its gravitational effects. We can try to measure the actual slowing down of the expansion of the universe and see if $-\ddot{R}$ is large enough to stop the expansion. When we do this, we are determining the current value of the deceleration parameter q_0. From (8.66) when applied to the present epoch, we get

$$q_o = -\ddot{R}(t_o)/R(t_0)H_o^2.$$

As we do not measure $R(t_o)$, we would like to express $R(t_0)$ in terms of $H(t_0)$ and $\dot{H}(t_0)$. Now $H(t) = \dot{R}(t)/R(t)$, from which we have

$$\dot{R}(t) = H(t)R(t).$$

Differentiating both sides with respect to time t gives

$$\ddot{R}(t) = H(t)\dot{R}(t) + \dot{H}(t)\, R(t).$$

Setting $t = t_0$, and remembering that $R(t_0) = 1$, this becomes

$$\ddot{R}(t_o) = H_o^2 + \dot{H}(t_o).$$

Substituting this back into the expression for q_0, we get

$$q_o = -\left[\dot{H}(t_o)/H_o^2 + 1\right]. \tag{8.68}$$

Equation (8.68) says that if we can measure $H(t_0)$ and $\dot{H}(t_0)$ we can determine q_0. How do we determine $H(t_0)$? We can take advantage of the fact that when we look deeper and deeper into space, we are also looking farther and farther back into time. Thus, if we can determine H for objects that are, say, five billion light-years away, we are really determining the value of H five billion years ago. If we include near and distant objects in a plot of Hubble's law, we will see deviations from a straight line. A sample result is shown in Fig. 8.7. On the horizontal axis, we plot

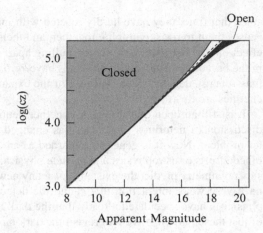

Fig. 8.7 Hubble's law with deviations for distant objects.

the brightness of a galaxy and on the vertical axis the redshift. The dashed line is an extrapolation of Hubble's law. If $q_o = 1/2$, the universe barely manages to expand forever; if $q_o > 1/2$, the universe is closed. If $q_o < 1/2$, the universe is open. In Fig. 8.6, the dark region is open, and the shaded region is closed. But measuring $H(t_o)$ is not easy. The difficulty comes in the methods for measuring distances to distant objects.

8.6 The Beginning, the End, and Time's Arrow

So far we have looked at the structure of the universe more or less from the point of view of space. We now give a brief discussion with the consideration of time, in particular, with time's arrow. Where does it fit in the overall nature of the universe?

The second law of thermodynamics leads us to speculate about the nature of time, and the way in which we can distinguish the past from the future. We observe in our environment that systems tend to evolve from ordered states to disordered state. As a measure of the disorderliness or lack of organization in a system, Rudolf Julius Emanuel Clausius introduced the concept of entropy: the greater the degree of disorder or randomness in a system, the greater is the entropy of the system. More-ordered systems have smaller entropy than less-ordered ones. He reformulated the second law of thermodynamics in terms of entropy

In any natural process taking place in an isolated system the entropy of the system either increases or remain constant. If this holds for all natural processes, the forward direction of time can then be unambiguously defined as the direction in which order diminishes and entropy increases. Because of this, entropy has often been described as "time's arrow," pointing from the past to the future.

This arrow of time based on the thermodynamic concept of entropy is often called the *thermodynamic arrow of time*. Can we also define a cosmological arrow of time?

Is there a possible connection between the arrows of time in thermodynamics and cosmology?

The cosmological arrow of time is the direction of time in which the universe is expanding. The answer to the question whether there is a possible connection with thermodynamic arrow of time rests on observation, which reveals that the universe is in a state of great order today, so, thermodynamically containing low entropy; and it is expanding toward increasing disorder and so entropy. If the two arrows of time point in the same direction, then the entropy of the universe was even lower in the past than today. In other words, the universe was created with a very high amount of order. Or did it get this way as an accident, a big "statistical fluctuation" that shows up once in a while, creating order in even the most disorderly system?

The past behavior of all three Friemann models is very similar: the scale factor of distances vanished ($R = 0$) at some moment in the finite past. The point $R = 0$ represents a singularity, like the center of a Schwarzschild black hole. This is partly correct. It is the time reversal of a black hole. The point $R = 0$ describes a situation where space-time came into existence; space expands at the beginning of time, and matter is exploding out of singularity. The singularity at the center of a black hole cannot be seen by us outside the horizon, whereas the singularity at the beginning of cosmological expansion is "naked," and in principle, we can look out into the universe back in time to see the creation. In practice, it is not possible to see back before about 10^5 years after the beginning of the expansion, because the cosmological material was opaque to radiation. Prior to that time, we have to rely on physics. General relativity, and possibly even the space-time description itself, breaks down at a sufficiently early stage in the expansion of the universe. It is estimated this occurs at a mere 10^{-43} second (the Planck time) after the initiation of the expansion. A quantum theory of gravity may help us to understand how the universe began.

Although it is not possible to continue known physics back as far as 10^{-43} seconds, it is possible to examine in great detail the processes occurring in the Friedmann universe in the early stages of the expansion. Some of the consequences of these processes are observable today, so that this simple model may be confronted with various observational data to test its reasonableness. It turns out that the Friedmann universe does remarkably well.

Like any physical system, the contents of the universe grow hot when compressed and cool when expanded. The famous redshift of spectral lines discovered by Hubble can be considered as a cooling of the light due to the cosmological expansion. It follows that during the early stages of the Big Bang, the universe was very hot. For that reason the contents of the universe during this epoch are usually referred to as the *primeval fireball*. None of the present structure, such as stars or galaxies which we observe in the universe today, could have existed in the fireball. Even atoms would have been smashed to pieces. The fireball at very early times should be envisaged as a fluid of all types of elementary particles (some as yet unknown in the laboratory), strongly interacting with each other, and in condition of thermal equilibrium. The $3\,K$ microwave background radiation is strong evidence

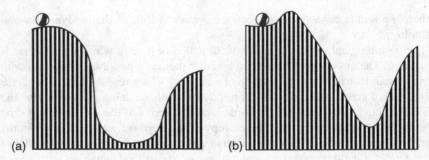

Fig. 8.8 (a) An unstable equilibrium. (b) A stable equilibrium.

that the universe was in a state of thermal equilibrium sometime after its creation, at which time the universe became transparent to radiation.

Now the important question is this: Once in equilibrium, how could the universe get itself into its present state of disequilibrium? How could a stable universe represented by the state of thermal equilibrium become unstable, as it is today?

The key part of the answer is that the primordial universe was in a state of local equilibrium, not in one of the most stable equilibrium. Figure 8.8 shows a ball in two different situations. In (a) the ball is totally unstable; the slightest disturbance will bring it to the valley of stability. After a few oscillations, it will settle at the bottom in total equilibrium. In (b) the ball is in local equilibrium; a big push to go over the barrier can bring it to the valley of stability. The primordial universe may have been like this ball, neither unstable nor stable, but metastable against small pushes. The radiation from this state is what we see today as the 3 degree background radiation.

Now, what enables the metastable equilibrium to become a state of disequilibrium? To answer this question it is necessary to explore the early universe and its subsequent evolution. We shall do this in some detail in Chapter 11. Briefly speaking, the disequlibrium of the universe is due to the changing conditions brought by its expansion and the cooling by radiation. As will be shown in Chapter 11, with the temperature falling rapidly from $10^{12} K$, the fireball began the so-called lepton era, with familiar protons, neutrons, and electrons as well as muons, neutrinos, and X-ray photons all jumbled together in equilibrium. The radiation was so hot that it could create electron-positron pairs. As the temperature dropped, first the muons disappeared, then the positrons. After about 10 seconds the temperature had fallen to a few billion degrees, and the principal interest centers on the remaining protons, neutrons, and electrons.

At this stage (which may be called the plasma era) the temperature has fallen low enough for the frantically moving neutrons and protons to start combining together to form helium and a few other light nuclei. Detailed calculations indicate that about a quarter of the protons got incorporated into helium nuclei, with a tiny proportion as deuterium and lithium. Thus about 25% of the nuclei that emerge from the fireball are helium, and the rest are hydrogen (single protons). This is remarkably close to the present observed abundances for these light elements. It is a valuable confirmation that the processes that occurred in the recombination era in the real universe were not far from that which the fireball model of the Friedmann universe suggests.

The plasma era continued for about 700,000 years, after which the temperature was down to 4000 K (a little cooler than the surface of the sun), and the electrons began to combine with the nuclei to form ordinary atoms. After this had occurred the way was clear for local condensations of matter to form under gravitational attraction. Clumps of gas were whirled up into clusters that slowly contracted to form galaxies and eventually stars and planets. The stellar interiors became a thermal nuclear factory to synthesize complex atomic nuclei, starting from hydrogen all the way to iron. When two light nuclei fuse, part of the total mass is converted to radiation energy (according to Einstein's formula $E = mc^2$) that then percolates slowly through the outer layers of the star and off into space. This process represented an entropy increase because the energy that was locked up in the nuclei was spread out into space. This pattern of disequilibrium through starlight and nucleosynthesis was repeated throughout the universe. The whole cosmos is in an unstable state, with vast cold emptiness punctuated sporadically by hot stars. These tremendous powerhouses of energy are continually pouring out light, heat, and other electromagnetic radiation in an attempt to redress the balance and restore thermal equilibrium. Therefore, stated in short, we can say that the disequilibrium of the universe is due to its expansion or the changing conditions brought by its expansion.

Our discussion above makes it clear that in the present phase of the universe, the cosmological arrow of time and the thermodynamic arrow of time point in the same direction; they are coupled. If the universe were open and continued to expand forever, then this coupling would never be broken. The thermodynamic arrow for such a universe predicts heat death, which is an end of time in a way. The cosmological arrow says just about the same thing: one by one the galaxies move so far away from ours that we won't be able to see any of them someday in the far future, and this time arrow also stops. Thus, there is an end of time of sorts in this kind of a model. It takes a very long time to achieve this, of course.

If the universe is closed, then in the expansion phase, the cosmological arrow and the thermodynamic arrow are in the same direction, but how about the contraction phase? Would the thermodynamic arrow reverse and disorder begin to decrease with time? The common belief is that the contracting phase is the time reverse of the expanding phase. This idea is attractive to many scientists because it would mean a nice symmetry between expanding and contracting phase. These scientists believe that the cosmological arrow is the dominant arrow, and it turns the thermodynamic arrow around. So the entropy actually decreases instead of increases (as reckoned by us). However, Hawking showed that this picture is wrong. His theory was based on the no boundary condition. What is this? The Big Bang was an explosion of space at the beginning of time. As time elapsed, space itself expanded. The universe has no center and no boundary or edge. So there would be no need to specify the boundary condition of space-time. Hawking and Jim Hartle worked out what conditions the universe must satisfy if space-time had no boundary. It is beyond the scope of this book to reproduce their theory here. But they showed that implied disorder in the no-boundary condition would in fact continue to increase during the contraction, and the thermodynamic arrow of time would not reverse.

8.7 An Accelerating Universe?

In the late 1990s two groups of astronomers found that the expansion of the universe over the past few billions years is speeding up rather than slowing down. This startling finding, if it holds up to further scrutiny, has far-reaching implications for the fate of the universe.

As we look at galaxies that are farther and farther away, we are looking farther and farther back in time. If the rate of expansion of the universe has changed over time, we should be able to see this change when we observe very distant (and thus very young) objects. To do this, we need to look at the universe when it was much younger, so we need an extremely bright standard candle that would be visible to huge distances. The best standard candle for these purposes is a special kind of supernova, called Type Ia (carbon detonation) supernovas. These objects are the brightest and can be seen from great distances; they all have about the same intrinsic brightness in nearby and distant galaxies. Moreover, their light curves obey a relationship between peak luminosity and the time it takes for the supernova to fade. This means that by measuring the light curve of a distant Type Ia supernova, we can determine its intrinsic brightness and, from that, its distance and how long ago the supernova occurred. From their redshifts, we can determine the rate of expansion of the universe in the past.

Type Ia supernovas occur about once every 300 years or so in a given galaxy, so in order to detect many supernovas astronomers need to monitor a huge number of galaxies. One of the teams monitors almost 100,000 galaxies. This is enough to detect dozens of supernovas during every observing period. Once supernovas are detected, their light curves are measured and their distances obtained. Using the Hubble Space Telescope or large telescopes such as the Keck telescope, astronomers also measure the redshifts of the galaxies in which the supernovas occur. Using these techniques, astronomers have found supernovas so distant that their light has been traveling toward us for 9.5 billion years– more than half the age of the universe.

If the universe has increased its rate of expansion, a supernova at a given redshift would lie farther away than expected, and so it would appear dimmer. In 1998 two teams of astronomers (S. Perlmutter, et al., of the Lawrence Berkeley Laboratory, and P. M. Garnavich and R. P. Kirshner of the Harvard-Smithsonian Center for Astrophysics) announced that distant type Ia supernovas are farther away than would be expected on the basis of their redshifts, and looked about 15% to 20% dimmer than expected. Thus the expansion of the universe over the past few billion years is speeding up instead of slowing down. Figure 8.9 is a schematic graph of the distance and recession velocity of distant galaxies based on observations by S. Perlmutter et al. Note that the distant galaxies lie below the line followed by nearby galaxies. This indicates that the expansion of the universe seems to have speeded up.

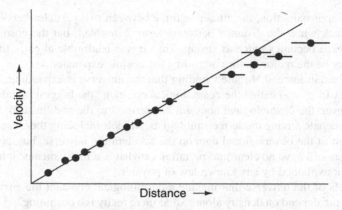

Fig. 8.9 A schematic graph of the distance and recession velocity of distant galaxies.

The measurements are difficult, and the results depend quite sensitively on just how "standard" the supernova luminosities really are. Nevertheless, most cosmologists have accepted these startling new results. For now, at least, the acceleration seems to be real.

The expansion of the universe could not speed up if the only contribution to the curvature of space is matter (normal or dark). According to the standard Big Bang, the universe has expanded ever since its explosive birth, but gravity has gradually slowed the expansion. Even if the universe grows forever, the theory predicts it should do so at a steadily decreasing rate.

The new finds are also in direct conflict with the widely accepted theory of inflation that explains why the structure of the universe looks the same in all directions. The theory of inflation also predicts that the cosmos has exactly the right density to bring expansion to an eventual halt.

What could cause an overall acceleration of the universe? Cosmologists do not know; it remains a deep mystery. But several possibilities have been suggested. Reconciling the standard Big Bang and the theory of inflation with endless expansion may require, some cosmologists think, the resurrection the so-called cosmological constant–an antigravity term in the equations of Einstein's general relativity, or some other exotic source of energy in the cosmos.

Einstein introduced the cosmological constant in 1917. When Einstein formulated the General Theory of Relativity in 1916, the universe was generally believed to be static, although this was simply a prejudice, rather than being founded on any observational facts. But when Einstein applied his field equations to a cosmic gas of the kind we have discussed, he found that if the density is not zero, the universe must necessarily be expanding or collapsing. To get his model static, he introduced into field equations a cosmological term that provides a repulsion mechanism. The coefficient of this new term, the number that determines how large an effect the term has, is called the cosmological constant, denoted by Λ (or λ). This repulsive force depends on distance differently than the attractive gravitational force. In the

Newtonian approximation, the attractive force between two particles becomes four times as weak when the distance between them is doubled, but the cosmological repulsive force becomes twice as strong. Thus it was negligible at early times, but today it may be the major factor controlling the cosmic expansion.

After Einstein learned Hubble's finding that the universe is expanding, he repudiated the Λ term, and called the cosmological constant "the biggest blunder of my life." However, the cosmological constant has refused to die and, instead, has generated new debate among modern cosmologists. Models including the cosmological constant can fit the observational data on the accelerating universe, but, at present, cosmologists still have no clear interpretation of what it actually means; it is neither required nor explained by any known law of physics.

In models of the universe that include a cosmological constant the curvature of space does not depend on density alone. So, if there really is a cosmological constant the future of the universe cannot be deduced just by determining the density of matter.

According to some astronomers, the mysterious cosmic field causing the universe to accelerate is neither matter nor radiation. It has become known as dark energy. If confirmed, the magnitude of the cosmic acceleration implies that the amount of dark energy in the universe may exceed the total mass-energy of matter (luminous and dark) by a substantial margin.

It is possible, though, that the two teams were fooled. Intervening dust could have made the supernovas look dimmer, or the more distant ones might have a slightly different composition than nearby supernovas, causing them to appear fainter.

Dust or composition differences could not mimic both deceleration at early times in the universe and acceleration at more recent times. By finding a large sample of supernovas that lie more than 10 billion light-years from Earth, astronomers might test whether cosmic acceleration is genuine. Thus, studying extremely distant supernovas is one of our best near-term bets.

8.8 The Cosmological Constant

As mentioned in the preceding section, to get a static universe Einstein introduced into field equations a cosmological term that provides a repulsion mechanism. How could this be done? We know that the covariant divergence of the Einstein tensor $G_{\mu\nu}$ ($G_{\mu\nu} = R_{\mu\nu} - g_{\mu\nu}R_\alpha^\alpha$) and the energy-momentum tensor $T_{\mu\nu}$ vanish identically; the metric tensor also has zero covariant divergence. Thus it is possible to write a modified set of field equations that are also consistent with the conservation laws:

$$R_{\mu\nu} - \frac{1}{2}g_{\mu\nu}R + \Lambda g_{\mu\nu} = -\frac{8\pi G}{c^4}T_{\mu\nu}$$

where Λ is the so-called cosmological constant.

Now, since $\Lambda c^4/8\pi G$ has the same dimension as the energy-momentum tensor, some physicists believe that the cosmological constant Λ is present even if the universe is totally devoid of matter and radiation, and that Λ can be thought of as the energy density of the vacuum:

$$\varepsilon_V = \frac{c^4}{8\pi G}\Lambda.$$

In models of the universe that include a cosmological constant Λ the curvature of space does not depend on mass density alone anymore; and the critical density ρ_c and the density parameter Ω_0 are given by:

$$\rho_c = \frac{3H_0^2 - \Lambda c^2}{8\pi G}, \quad \Omega_0 = \frac{8\pi G\rho_0}{3H_0^2 - \Lambda c^2}.$$

From this follows an estimate for Λ, based on the condition that the critical density must not be zero:

$$\Lambda \leq \frac{3H_0^2}{c^2} \approx 3.5 \times 10^{-56}\, cm^{-2}.$$

Note that the square root of the reciprocal of Λ is a distance. It can, in principle, be determined from observation. In the presence of a nonzero Λ the future of the universe cannot be deduced just by determining the density of matter.

The cosmological constant Λ has also experienced a revival through quantum field theories. In quantum field theories the vacuum is defined as the state of lowest energy. Anything contributing in some form to the vacuum energy density also provides a contribution to the cosmological constant. There exist, in principle, three different contributions:

$$\Lambda_{tot} = \Lambda_{ein} + \Lambda_{quan} + \Lambda_{int}$$

where Λ_{ein} is the one introduced by Einstein, often called the bare cosmological constant, the value the cosmological constant would have if none of the particles existed and if the only force were gravity; Λ_{quan} is due to quantum fluctuation; and Λ_{int} is similarly to Λ_{quan}, due to possible particles and interactions, such as Higgs field and Higgs bosons.

We can neglect Λ_{int} at the moment, as we don't know it very well. Let us consider Λ_{quan}. Quantum fluctuations manifest themselves as pairs of virtual particles that appear spontaneously, briefly interact, and then disappear. Although virtual particles cannot be detected by a casual glance at empty space, they have measurable impacts on physics, and in particular they contribute to the vacuum energy density. The contribution made by vacuum fluctuations in the standard model depends in a complicated way on the masses and interaction strengths of all the known particles. As a simple example, we consider a quantum harmonic oscillator. Its eigenvalues are given by

$$E_n = \left(n + \frac{1}{2}\right)\hbar\omega, \quad n = 0, 1, 2.$$

The vacuum ($n = 0$) has a finite amount of energy (zero point energy). A relativistic field can be considered as a sum of harmonic oscillators of all possible frequencies ω. For the simple case of a scalar field with mass m, the vacuum energy is given by a sum:

$$E_0 = \sum_j \frac{1}{2}\hbar\omega_i.$$

This summation can be rewritten as integration by putting the system in a box of volume L^3 and then letting $L \to \infty$. If we impose periodic boundary condition, the above summation then becomes

$$E_0 = \frac{1}{2}L^3 \int \frac{d^3k}{(2\pi)^3}\omega_k$$

where we have set $\hbar = 1$, $k = 2\pi/\lambda$ corresponding to the wave vector. The integration can be carried out if we use the relation

$$\omega_k^2 = k^2 + m^2$$

and a maximum cut-off frequency $k_{max} \gg m$. The result is

$$\rho_V = \lim_{L\to\infty} \frac{E_0}{L^3} = \int_0^{k_{max}} \frac{4\pi k^2}{(2\pi)^3}dk\frac{1}{2}\sqrt{k^2 + m^2} = \frac{k_{max}^4}{16\pi^2}.$$

The general relativity is valid up to the Planck scale. Setting $k_{max} = l_p$ we get

$$\rho_V \approx 10^{92}\,\text{g}\cdot\text{cm}^{-3},$$

which is 121 orders of magnitude above the experimental value. Obviously it is not correct. In order to estimate the contribution of a single particle species, we assume the virtual particles produced take up for a short time their Compton volume L_c^3, L_c is the Compton wavelength ($L_c = \hbar/mc$); then

$$\rho_V = \frac{m}{L_c^3} = \frac{c^2 m^4}{\hbar^3}.$$

Inserting for m the mass of, for example, the u and d quarks, their contribution to the vacuum energy alone would produce an effect on a scale of about $1/(1\,\text{km}^2)$. Contribution of the W and Z bosons would be noticeable on a scale of $1/(20\,\text{cm}^2)$. This means that effects of the curvature of space would appear on scales of meters to kilometers. Obviously this is not what we see.

The geometric structure of the universe is extremely sensitive to the value of the vacuum energy density. If the vacuum energy density, or equivalently the cosmological constant, were as large as theories of elementary particles suggest, the universe in which we live would be dramatically different, with properties we would find both bizarre and unsettling. Physicists don't know what is wrong with the theories at present. Many different attempts, such as supersymmetry, fluctuations in the topology of the geometry of space-time, and others, are under consideration as possible solutions for the Λ problem. It is beyond the scope of this book to discuss these attempts.

8.9 Problems

8.1. Check the expressions for the Ricci tensors and the Einstein tensors for the Robertson-Walker line element (i.e., [8–3], [8–4], [8–5] and [8–6]).

8.2. A galaxy is observed with a redshift of 0.69. How long did light take to travel from the galaxy to us if we assume that we are in the Einstein-de Sitter universe?

$$(H_o = 100 \, \text{km/s Mpc})$$

8.3. From the equation $T^{\mu\nu}_{;\mu} = 0$ obtain Equation (15) and deduce that, if $p = 0$, then $\rho \propto 1/R^3$.

8.4. Show that in the presence of Einstein's λ term, Equations (7) and (8) are modified to the following:

$$2\frac{\ddot{R}}{R} + \frac{\dot{R}^2 + kc^2}{R^2} - c^2 = \frac{8\pi G}{c^2} T_1^1$$

and

$$\frac{\dot{R}^2 + kc^2}{c^2} - \frac{1}{3}c^2 = \frac{8\pi G}{3c^2} T_o^o.$$

8.5. Show that Newtonian gravity does not admit a cosmology that is isotropic, homogeneous, and static.

8.6. Show that an empty (containing no matter) isotropic space-time is a flat Minkowski space.

References

Hawking S (1988) *A Brief History of Time.* (Bantam Books, New York)

Landau LD, Lifshitz EM (1975) *The Classical Theory of Fields.* (Pergamon Press, New York)

Rindler W (1977) *Essential Relativity.* (Springer-Verlag, New York)

Schwarzschild B (1998) Very distant supernova suggest that the cosmic expansion is speeding up. Phys Today, July

Chapter 9
Particles, Forces, and Unification of Forces

For the following chapters, you will need some knowledge of modern particle physics. The early universe was a microscopic world, the subjects within it interacting by the forces that particle physicists study. As such, it is a very exciting subject to explore for anyone who is curious and has an imagination.

At the beginning of the 1930s, it looked as if physicists had a very good grasp of what the world was made of. There were only four particles—the electron, the proton, the neutron, and the photon—and there were two fundamental forces: gravity and electromagnetism. But soon the world began to appear to be a more complex place. During the next 20 years physicists identified as many elementary or fundamental particles (particles without internal structure) as there are different chemical elements. Trying to bring some order to this proliferation of particles, Professor Murray Gell-Mann introduced the idea of quarks as fundamental particles.

Today the standard model of particle physics describes our current picture of matter and the interactions responsible for all processes. Hundreds of subatomic particles and their properties are now understood in terms of 6 quarks and 6 leptons, from which all matter is made; the interactions act through the exchange of carrier (or messenger) particles. Photon, for example, is the carrier particle for electromagnetic interaction.

9.1 Particles

9.1.1 Spin

Particles have many attributes that are vital to their interactions, such as mass and electric charge. All particles also have another property, called "spin," a property that is expressed in terms of a unit called Planck's constant h. Because the unit for spin is always $\hbar(= h/2\pi)$, it is usually omitted. In quantum theory things come in discrete amounts; particles can have 0 spin, spin 1/2, spin 1, spin 3/2, etc.

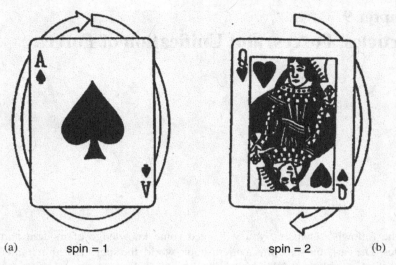

Fig. 9.1 (a) The particle of spin 1 is like an arrow. (b) The particle of spin 2 is like a double-headed arrow.

Many textbooks on general physics and modern physics suggest that one way of thinking of spin is to imagine the particles as little tops spinning about an axis. But this can be misleading, because quantum mechanics tells us that the particles do not have any well-defined axis. What the spin of a particle really tells us is what the particle looks like from different directions. A particle of spin 0 is like a dot: it looks the same from every direction. On the other hand, a particle of spin 1 is like an arrow: it looks different from different directions (Fig. 9.1a). But if we turn it around a complete revolution (360 degrees) the particle will look the same. A particle of spin 2 is like a double-headed arrow (Fig. 9.1b); it looks the same if we turn it around half a revolution (180 degrees). Higher-spin particles look the same if we turn them through smaller fractions of a complete revolution. All this seems fairly straightforward, but the remarkable fact is that there are particles that do not look the same if we turn them through just one revolution; we have to turn them through two complete revolutions! Such a particle is said to have spin $1/2\hbar$.

A spin can make itself known through many effects. Physicists sort particles, according to their spin, into two groups: fermions and bosons. They are named, respectively, after the Italian-American physicist Enrico Fermi (1901–1954), and the Indian physicist Satyendra Nath Bose (1894–1974).

9.1.2 Fermions

Particles with half-integer spin, $1/2\hbar$, $3/2\hbar$, etc., are called *fermions*. They obey Pauli's exclusion principle (no two fermions in the same system can occupy the same "state" at the same time), and they are matter particles. That is, all matter in the universe appears to be composed at some level of constituent fermions.

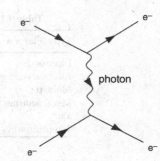

Fig. 9.2 Feynman diagram for e⁻e⁻ encounter.

9.1.3 Bosons

Bosons are particles with integer spins, 0, 1, 2, 3, etc. They do not obey the exclusion principle; consequently, bosons all tend to occupy the state of the system with the lowest possible energy. The fundamental forces are transmitted by exchanges of bosons. So bosons are carrier particles, not matter particles. For example, the electromagnetic interaction between two charged particles is mediated or transmitted by the exchange of a photon, as shown in Fig. 9.2; the photon is a boson with spin $0\hbar$. The weak force is mediated by 3 quanta of the weak field: W^+, Z^0, W^-. They are (intermediate vector) bosons with spin $1\hbar$. The carrier particles of the strong field are called gluons; there are 8 gluons, all having spin $2\hbar$.

Figure 9.2 is a Feynman diagram, which is a space-time diagram that provides a useful way of visualizing interactions between particles.

There is one other boson that has been predicted, but not yet detected, that seems to be necessary in quantum field theory to explain why the $W^\pm$ and Z^0 have large masses, yet the photon has no mass (or a negligibly small mass). These not-yet-detected bosons are called Higgs bosons or Higgs particles, after Peter Higgs, who first proposed them. We know very little about them.

9.1.4 Hadrons and Leptons

Hadrons are composite particles made up of strongly interacting constituents, the quarks. There are two classes of hadrons–baryons and mesons–and their properties are quite different. Baryons are heavy particles, and fermions, protons, and neutrons are baryons with spin $1/2\hbar$. Mesons are medium-weight particles and are bosons (integral spin). Pions and kaons are mesons. All baryons have masses at least as large as the proton and half-integral spin (fermions). Baryons heavier than two nucleons are called hyperons; they are all unstable. The proton is the only stable baryon, but some theories predict that it is also unstable with a lifetime greater than 10^{30} years.

Hadrons $\begin{cases} \text{Baryons (fermions, heavy particles)} : p, n, \text{etc} \\ \text{Mesons (bosons, medium-weight particles)} : \pi^\pm, \pi^0, W^\pm, Z^0, \text{etc.} \end{cases}$

Table 9.1 Leptons (Fermions, spin = $1/2\hbar$)

Type (Flavor)	Mass (GeV/c^2)	Electric charge
Electron e	5.1×10^{-4}	-1
Electron neutrino v_e	$<1 \times 10^{-8}$	0
Muon μ	0.106	-1
Muon neutrino v_μ	$<1.7 \times 10^{-8}$	0
Tau τ	1.777	-1
Tau neutrino v_τ	$<1.8 \times 10^{-2}$	0

Particles that do not feel the strong interactions are called leptons; they are light particles–fermions with spin $1/2\hbar$. Leptons are the simplest elementary particles. They appear to be point-like and seem to be truly fundamental with no internal structure. There are only six known leptons: electron, muon, tau, and their associated neutrinos, grouped into three families or generations.

First generation: electron *e* and electron neutrino v_e
Second generation: Muon μ and muon neutrino v_μ
Third generation: Tau τ and tau neutrino v_τ

In all known weak interaction processes, the v_e and electron are paired; the v_μ is paired with μ, and the v_τ is paired with τ. This pairing appears to be a basic or fundamental part of nature's pattern.

Note that the masses of leptons in Table 9.1 are expressed in units of GeV/c^2. Particle physicists measure energy in units of electronvolts eV, megaelectrovolts MeV, or gigaelectronvolts GeV. Hence, because $E = mc^2$, the unit of mass is MeV/c^2 or GeV/c^2.

Particle physicists have accumulated enough evidence to suggest the following rule: In any reaction, the total number of particles from each lepton generation must be the same before the reaction as after (Lepton conservation).

The neutrino, an exotic particle, merits a little special attention. W. Pauli postulated its existence in 1931 in order to explain neutron decay. If a neutron decayed into an electron and proton, then they would move off along a straight line. In practice they are seen to move off at an angle to one another. Pauli believed this is due to a third invisible particle being produced, the neutrino (the little neutral one). The existence of the neutrino also explains the otherwise anomalous energy in the neutron decay. C. Cowan and F. Reines finally detected the neutrino in 1956. Reines was awarded the Noble Prize in 1995 (Cowan died in 1995).

There are three types of neutrinos: v_e, v_μ, v_τ. Recent experiments at the SLA and CERN laboratories prove that only three low-mass neutrinos exist, but don't exclude the possibility that extraordinarily massive neutrinos can also exist.

The neutrinos have no electric or strong charge (to be explained later). They interact so little that they do not form compound objects. They are not affected by electromagnetic or strong forces, but by the weak and gravitational forces. Gravity would be extremely weak for an individual neutrino. If there are enough neutrinos with mass in the universe, their combined gravity might stop the expansion of the universe.

9.1.5 Quarks

Quarks, like leptons, are fundamental particles. The biggest difference between quarks and leptons is that quarks are affected by strong forces that bind them into composite particles, such as protons and neutrons. In contrast, leptons do not experience strong interaction and can exist as separate objects. There are six quarks; all are fermions with spin $1/2\hbar$. They also group into three generations or families. The u and d quarks are paired as first generation, c and s as the second generation, and t and b as the third generation:

Each quark has its own antiquark. All the properties of a given antiquark (except mass) are the negative of those for the corresponding quark.

According to the standard model of particle physics, matter is made up of the 12 fundamental particles (6 leptons, 6 quarks) listed in Tables 9.1 and 9.2. Twelve particles—that is all that makes up the matter of the entire world. These 12 particles are the fundamental particles of everything. Most matter around us is made up of the u and d quarks because they are the only stable quarks.

You may already have noticed that there is one familiar thing and one surprise in Tables 9.1 and 9.2. The familiar thing is the electron, which is one of the constituents of the atom and the particle that is responsible for the electric current in wires. The surprise is that the proton and the neutron are missing in the tables. We were told in general physics that every atom has a small, heavy, positively charged nucleus, in addition to the orbiting negative electrons, and the nucleus is composed of positively charged proton and neutral neutrons. What is the answer to this vanishing act of the proton and neutron? We now know that they are not fundamental; they are composed of quarks. The proton is composed of two up quarks and one down quark. And the neutron is composed of two down quarks and one up quark: $p \equiv uud$ charge $= 2/3 + 2/3 + (-1/3) = 1$; and $n \equiv udd$ charge $= (-1/3) + (-1/3) + 2/3 = 0$.

The quarks and antiquarks can combine only in ways that produce integral charges. The standard model of particle physics suggests that baryons are composed of three quarks, and antibaryons consists of three antiquarks. Mesons are made up of a quark and an antiquark.

The standard model of strong interaction implies that quarks are "colored": red, blue, and green. Each quark carries one of these three types of color charge (or strong charge). Each antiquark has one of the three complementary color (also called anticolor) charges. The name "color" has nothing to do with the colors of visible

Table 9.2 Quarks (Fermions, spin $= 1/2$)

Type (Flavor)	Mass (GeV/c^2)	Electric Charge
u up	0.005	2/3
d down	0.01	−1/3
c charm	1.5	2/3
s strange	0.2	−1/3
t top	175	2/3
b bottom	4.78	−1/3

Table 9.3 Sample Baryons (Fermionic Hadrons)[1]

Symbol	Name	Quark Content	Electric Charge	Mass GeV/c^2	Spin
P	Proton	Uud	1	0.938	1/2
$\bar{P}$	Antiproton	$\bar{u}\bar{u}\bar{d}$	-1	0.938	1/2
N	Neutron	Udd	0	0.940	1/2
Λ	Lambda	Uds	0	1.116	1/2
Ω^-	Omega	Sss	-1	1.672	3/2

[1](Baryons: qqq; Antibaryons: $\bar{q}\bar{q}\bar{q}$) q stands for quark and $\bar{q}$ for antiquark.

Table 9.4 Sample Mesons (Bosonic Hadrons, q)

Symbol	Name	Quark Content	Electric Charge	Mass GeV/c^2	Spin
π^+	Pion	$u\bar{d}$	$+1$	0.140	0
K^-	Kaon	$s\bar{u}$	-1	0.494	0
ρ^+	Rho	$u\bar{d}$	$+1$	0.770	1
D^+	D$^+$	$c\bar{d}$	$+1$	1.869	0
η_c	Eta-c	$c\bar{c}$	0	2.980	0

light; it is simply a cute name that was used. We will explain in next section that why the concept of color charges was introduced.

In electromagnetism, electric charge can be positive and negative. When the number of the positive and negative charges of an object are equal, it is electric neutral. In a sense, this idea is used in the theory of strong interaction; color charges determine if a system is color neutral. But for quarks there are three types of strong charges (colors) and three opposite strong charges (anticolors). It is this property of combining three different color charges to produce a color-neutral (colorless) object that suggested the term color charge for the strong charges of the quarks. We will see in next section how any stable hadron must be a colorless combination of quarks.

9.1.6 Quark Colors

You may already have noticed two problems with the quark theory. One is that there was no explanation for why the only allowed combinations are three quarks or one quark and one antiquark. The other is why we have not been able to detect free quarks. There was another problem. Quarks have the same spins as electrons. Therefore, they should obey Pauli's exclusion principle. However, some particles are observed that are clearly combinations of three identical quarks (uuu), for example, and all in the ground state. This is a violation of the exclusion principle. To get out of this problem, Greenberg and Nambu suggested that there is a strong force charge comparable to the electric charge that a particle has to experience the electromagnetic force. Electrical charge comes in two varieties: positive and negative. But the strong force charge comes in three types that could be different for each

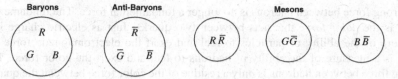

Fig. 9.3 Quark color combinations.

of the three quarks in a baryon. Greenberg and Nambu named this property *color charges* or *colors* for simplicity: red (R), green (G), and blue (B). To distinguish them from colors, we call the six quark types (u, d, s, c, t, b) *flavors*. Each flavor of quark comes in each of the three colors. The six antiquarks come in corresponding anticolors.

The standard model theory of strong interaction implies that any stable hadron must be a colorless combination of quarks. The rules for combining color are the following:

(1) A color and its anticolor cancel each other out. We call this colorless (or white).
(2) All three colors or all three anticolors in combination also cancel each other out and give colorless.

Thus, a baryon must contain a red, blue, and green quark. For example, a proton could be u(R)u(B)d(G), or u(R)u(G)d(B), or u(G)u(B)d(R), etc. A meson is a colorless combination of color and anticolor. For example, a π^+ could be red-antired $u(R)\overline{d}(\overline{R})$, green-antigreen $u(G)\overline{d}(\overline{G})$, or a blue-antiblue $u(B)\overline{d}(\overline{B})$. We illustrate the rule of combining quarks in Fig. 9.3. Three quarks are needed to make baryons and antibaryons. For the baryons we need one of each color; for the antibaryons we need one of each anticolor. The three circles at the right show how to combine a quark and an antiquark of a color and an anticolor pair to produce a meson. Once the right color combinations are present, flavor combinations are allowed.

Note that hadrons are combinations of color-charged quarks, but hadrons themselves do not have color charge. They are colorless in the same way that atoms are electrically neutral, even though they contain electrons and protons.

9.1.7 Quark Confinement

The introduction of quark color has also solved the quark confinement problem. Physicists now believe that free quarks cannot be observed; they can only exist within hadrons. This is known as *quark confinement*. A free quark would not be colorless, and so is not allowed. In other word, quarks are always confined within nucleons (protons and neutrons).

The properties involving quark color are not just a set of *ad hoc* rules. They have actually been derived from a mathematical theory called *QCD* (the *quantum chromodynamics*), developed in analog to QED quantum electrodynamics. In QCD, the strong force between hadrons is no longer a fundamental force. The fundamental force is the *color force* that acts between two quarks. Just as electric charge is a measure of the ability of particles to feel and exert the electromagnetic force, so color is a measure of the ability of hadrons to feel and exert the color force. The strong force between hadrons is only a residue of the color force between the quarks within the hadrons. By analogy, in the 19th century physicists believed that the force between neutral molecules called the van der Waals force was a fundamental force. After the development of the atomic theory in the early 20th century, it became clear that this was nothing more than the residual force between the electrons and protons within the molecules.

We have seen that quantum theories of forces involve carrier particles. QCD is no different. In fact, the mathematical theory predicts the existence of a group of eight particles carrying the strong force, called *gluons*. Gluons are massless and have no electric charge. There is a major difference between QED and QCD. The photons that carry the electromagnetic force have no electric charge themselves; the gluons that carry the color force have color charges represented by suitable combinations of the colors R, G, or B. When a quark emits or absorbs a gluon, its color can be changed. For example, an R quark emitting a R-G gluon becomes a G quark. This is a complicated effect, and that is why chromodynamics is more complex than quantum electrodynamics. The detailed calculations that have characterized the success of QED have not yet been possible for QCD.

In electromagnetic interaction, the active particles (electrons, protons, and so forth) and their carriers (photons) can exist in their free state. However, neither quarks nor gluons are ever seen as free particles. Is this a fatal blow to the theory? The answer is no. One possible way out comes from the curious phenomenon of the polarization of the vacuum that is allowed by Heisenberg's uncertainty principle. The importance of vacuum polarization (or vacuum fluctuations) in electromagnetic processes has long been experimentally confirmed (Lamb shift of the spectral lines of the hydrogen atom). The vacuum polarization comes into play for the quark in a little more complex way. A sea of virtual gluons as well as pairs of quarks and antiquarks surrounds the quark q. These virtual gluons, unlike photons that have no electric charge, have a color charge; rather than shielding q, they reinforce and extend q's charge. The net effect of the vacuum polarization is to amplify the intensity of the strong interaction at a distance and to decrease it locally. At very short distances the intensity of the strong interaction asymptotically approaches zero, and quarks become independent of each other at very close distances. However, if we try to separate them from each other, we have to struggle against an interaction that becomes stronger and stronger, and thereby endows the quarks with more and more energy. A final stage will be reached at which this energy, rather than being used to separate the quarks, instead is used in the creation of quark and antiquark pairs (Fig. 9.4), which in turn form new nucleons and pions. Thus, it is impossible to observe free quarks.

Before we move on to next section, let us summarize in Fig. 9.5 the classification of elementary particles:

Note that the carriers of mass (matter particles) are all fermions, and the carriers of forces (carrier particles or exchange particles) are all bosons.

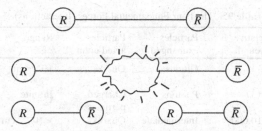

Fig. 9.4 One picture of quark confinement.

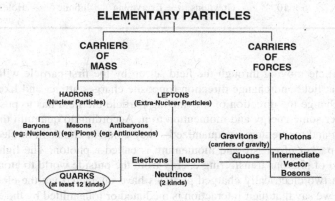

Fig. 9.5 Left: The scattering of muon neutrinos from an electron via a Z^0. Right: The scattering of electron neutrino from an electron via a W^+.

9.2 Fundamental Interactions and Conservation Laws

Along with the quest for the fundamental particles, physicists are also trying to understand the forces with which the particles interact. The concepts of forces and particles are intimately tied together. Without forces, particles would have no meaning, since we would have no way of detecting them.

There are many forces that occur in different situations. As an example, friction is always there when two objects slide against each other. The theory of how frictional forces arise is very complex, but in essence they are due to the electromagnetic forces between the atoms of one object and those of another. A fundamental force is one that does not arise out of a more basic force, just as a fundamental particle is one that is not composed of any other particle. We now recognize four forces as being sufficiently distinct and basic to be called fundamental forces: gravity, electromagnetic, the weak force, and the strong force (summarized in Table 9.5).

Gravity is incredibly weak, yet it binds us to the Earth, keeps the planets in orbit around the sun, and determines the fate of our universe. After gravity, the effects of the electromagnetic force are the next most familiar to us. Any particle with electric charge sets up an electromagnetic field throughout space. The intensity of the field decreases farther away from the particle (as the square of the distance). Any other

Table 9.5 The Four Fundamental Forces (Interactions)

Force	Relative strength	Particles exchanged	Particles acted upon	Range	Example
Strong	1	Gluons	Quarks	10^{-13} cm	Holds nuclei together
Electromagnetic	1/137	Photons	Charged particles	Infinite	Holds atoms together
Weak	1/10,000	Intermediate vector bosons	Quarks, electrons, neutrinos	$<10^{-14}$ cm	Radioactive decay
Gravity	6×10^{-39}	Gravitons	Everything	Infinite	Holds the solar system together

charged particle moving through the field set up by the first particle will feel the effect of that field and change direction (opposite charges attract and like charges repel). To change the direction of the moving particle the field has to push or pull it, or transfer some energy and momentum to it. According to quantum theory, the field energy and momentum are quantized—they come in chunks. For electromagnetism the packet of energy and momentum is called a photon. The light we see is composed of photons transferring energy from the outside world to atoms in our eyes. When two electrically charged particles have interacted via the electromagnetic force, we say that their interaction is mediated or transmitted by the exchange of a photon, as shown in Fig. 9.2. Just as the words *force* and *interaction* can be used interchangeably, so can *transmitted* and *mediated*. The photon, and any other particle that transmits a force because it is the quantum of a field, is called a "boson." The quantum theory of electromagnetism is called *quantum electrodynamics* (QED). Because the range of the electromagnetic interaction is infinite, the photon is massless. If the exchange particle has a high mass, it will be difficult to produce and exchange it over a large distance, so the force that it transmits will have a short range. We can understand this in terms of Heisenberg's uncertainty principle. The relation between the mass of the exchange particle and the time interval of its existence (lifetime) is

$$\Delta E \cdot \Delta t \geq \hbar \quad (\hbar = h/2\pi)$$

where ΔE = the temporary fluctuation in energy of the system needed for the rest mass m of the exchange particle, and Δt = the time interval of its existence.

The range r of the interaction is given by

$$r = c\Delta t = c\hbar/\Delta E.$$

Now,

$$\Delta E = mc^2.$$

Thus,

$$r = \frac{c\hbar}{mc^2} = \frac{\hbar}{mc}.$$

The range r is inversely proportional to m. For electromagnetic interaction $r \to \infty$, and accordingly $m = 0$. That is, the photon is massless.

QED has been tested in many ways to a very high accuracy, and is a very successful theory. It is speculated that the gravitational interaction is carried by a massless particle, called the graviton, since gravity has the same long-range behavior as electromagnetism. But no gravitons have ever been detected. This theoretical framework is still being developed. We will see later that the absence of a quantum-mechanical theory of gravity provides the limitation on how far back we can go in probing the Big Bang.

The weak force has no easily recognizable effects in the everyday world, but it is nevertheless of great importance. Just as the photon mediates the electromagnetic force, the weak force is mediated by quanta of the weak field. The amount of weak charge that particles can carry comes with more possible values than electric charge; three bosons, called W^+, Z^0, and W^- were predicted in the 1960s by Glashow, Salam, and Weinberg. They were finally discovered in 1983 at CERN, and give a firm confirmation of the electroweak theory of Glashow, Salam, and Weinberg. The masses of W and Z particles are much greater than the proton mass.

Figure 9.6a shows the scattering of a ν_μ (muon neutrino) from an electron, which involves the exchange of a Z^0 (this exchange is called a neutral current interaction). The scattering of an electron neutrino from an electron may also occur via a Z^0, but it may also involve the exchange of W^+ (a charge current interaction).

The weak force is actually weak not because its strength is less than the electromagnetic force, but because the range of the force is very short. So it is very unlikely that the two particles are close enough together for one to feel the other's weak force. The range is very short because the W and Z bosons that mediate the weak force are so heavy that it is hard for the two particles to exchange them. The weak field around a lepton or quark extends a much shorter distance than its electromagnetic field.

Table 9.6 Electroweak Force Carrier (Spin 1 $h/2\pi$)

Name	Mass (GeV/c^2)	Electric Charge
γ photon	0	0
W^-	80.6	-1
W^+	80.6	$+1$
Z^0	91.16	0

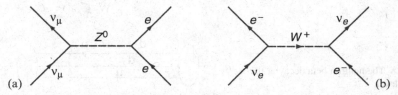

(a)　　　　　　　　　　　　　　　　　　(b)

Fig. 9.6 (a) The effect of the weak force on the leptons. (b) The effect of the weak force on the quarks.

All particles with a weak charge (all quarks, leptons, and W^+, Z^0, and W) feel the weak force; only photons and gluons do not. The weak force can change one lepton into other lepton within the same family. The weak force can also turn one quark into another within the same family or generation: (u, d), (c, s), and (t, b). We illustrate this in Table 9.7.

We see that a u quark can turn into a d quark by emitting a W^+. It is almost true to say that the weak force cannot cross quark generations. It can, but with a much reduced effect. Figure 9.7 illustrates this, with the heavy double arrow standing for easy transition and light double arrow for a more difficult transition:

The most dramatic effect of the weak force is that it makes all the quarks but the lightest one (u), and all the electrically charged leptons but the lightest one (e), unstable. Because of this instability the heavier quarks decay into the up quark and leptons decay into the electron and neutrinos (the strange quark and the muon decay in about one-millionth of a second, the other quarks and leptons even faster). Figure 9.8 describes the neutron beta decay: $n \rightarrow p + e + \bar{v}_e$:

Figure 9.8 also shows that at the quark level the decay is

$$d \rightarrow \quad u + e^- + \bar{v}_e.$$

Table 9.7 All Particles with Weak Charge Feel Weak Force

	e	v_e		u	D
E	Z^0	W^-	U	Z^0	W^+
v_e	W^+	Z^0	D	W^-	Z^0

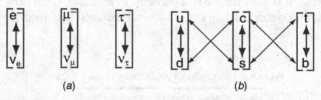

(a)   (b)

Fig. 9.7 (a) The effect for the weak force on the leptons. (b) The effect of the weak force on the quarks.

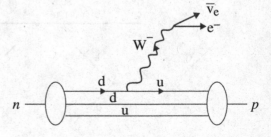

Fig. 9.8 The neutron beta decay at the quark level.

All processes in which the quark or lepton type changes are now understood to be weak interaction processes. In a universe with no weak interactions many different kinds of atoms and nuclei would have existed, leading to many different phenomena.

As we just saw, the strong force between hadrons (i.e., protons and neutrons) is a residue of the color force that only acts between quarks. The color force is mediated by the exchange of gluons, quanta of the color field set up by the quarks. (Only the quarks and the gluons themselves carry the color charge; the leptons and γ, $W+$, Z^0, and W^- do not). The weak force requires three bosons to transmit the weak interaction between particles of different weak charge. The color force requires eight gluons, each with a different color charge, to mediate all effects of the color force. We illustrate this in Table 9.8 for u quarks.

We see that an uB quark can turn into an uR quark by emitting a $(B\text{-}R)$ gluon. Figure 9.9 shows the exchange of a gluon $(B\overline{R})$ between a quark having color R and a quark having color B; the colors of the quarks are changed in the interaction.

The color force binds quarks into protons and neutrons. The protons and neutrons have no net color charge, just as atoms have no net electric charge. But just as there is residue force between the electrons and protons within the molecules that gives rise to the van der Waals force between molecules, here there is a leakage of color force outside the proton and neutron. This residual color force is the strong force that binds the protons and neutrons into the nuclei of the chemical elements.

The most widely accepted theory of elementary particle physics at present includes the existence of the standard model. It is a combination of the quark model of particle structure, the electroweak theory (a unified theory of electromagnetic and weak interactions), and the strong-interaction analog of quantum electrodynamics (QED) called quantum chromodynamics (QCD). Its details are too complicated to present here, but its general ingredients have already been presented above.

Although the standard model has been successful in particle physics, it doesn't answer all the questions. For example, it is not by itself able to predict the particle masses but must rely on guesses for many parameters in order to calculate masses. It

Table 9.8 Color Forced Mediated by Exchange of Gluons

	uR	UG	UB
UR	g(R-R)	G(R-G)	g((R-B)
UG	g(G-R)	G(G-G)	g(G-B)
UB	g(B-R)	G(B-G)	g(B-B)

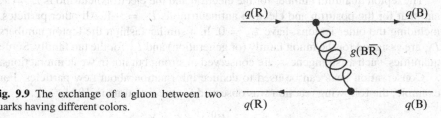

Fig. 9.9 The exchange of a gluon between two quarks having different colors.

$$p \ p \longrightarrow Z^0 \ Z^0$$

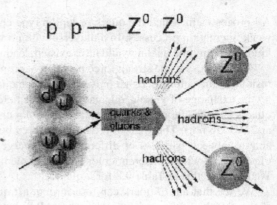

Fig. 9.10 A possible production of a Higgs boson by a proton-proton collision.

seems necessary to have one more type of interaction or field, which is essential for generating mass. This is the postulated Higgs field. Like magnetic or gravitational fields, the Higgs field permeates all space. But unlike them, its interaction does not cause a force on particles; rather it gives the particles their mass. Photons are massless because they do not interact with the Higgs field, while the W and Z do interact with the Higgs field. The carrier particle associated with the Higgs field is called the Higgs boson. It is a spin-zero particle and could have a large mass. Such a particle has not yet been observed. As it is an essential feature of the Standard Model of particle physics, a major goal of many experiments is to find this Higgs boson. Figure 9.10 shows a possible production of a Higgs boson by a proton-proton collision. The Higgs boson decays into two Z^0 bosons. The Z^0 is very useful because, unlike the photon, it interacts directly with both leptons and quarks. The Z^0 decays into a variety of particle and antiparticle pairs, each of which may allow physicists to unravel one more clues. At this time considerable data are being collected on the Z^0 and physicists hope to find some answers in the near future.

Conservation laws play an essential role in all interactions. In general conservation laws are associated with symmetries of the Hamiltonian of a physical system. In addition to energy, momentum, electric charge, and angular momentum, all interactions obey the following conservation laws:

(1) *The baryon number is conserved.* All baryons have baryon quantum number $B = 1$; all antibaryons have $B = -1$.
(2) *The lepton number for each family (or generation) is independently conserved.*

The lepton quantum number for the electron and the electron neutrino is $L_e = 1$, and that for the positron and electron antineutrino is $L_e = -1$. All other particles, including the other leptons, have $L_e = 0$. In a similar fashion the lepton numbers L_μ are assigned for the muon family (or generation) and L_τ for the tau family. Some quantities, such as strangeness, are conserved in strong but not in weak interactions.

Conservation laws can be used to deduce information about new particles. For example, the following reaction was observed for the first time in 1964:

$$K^- + p \rightarrow K^0 + K^+ + X.$$

What can we learn about the new particle X from this reaction? Let us apply the conservation laws. Charge conservation requires that X must carry a negative charge. Conservation of baryon number tells us that X is a baryon; kaons are mesons and proton is a baryon ($B = 1$), so X must have baryon number $B = 1$. This tells us that X has a qqq combination. To find out the specific combination, let us see the quark contents of the particles before and after the reaction:

$$K^- + p \rightarrow K^0 + K^+ + X$$
$$s \quad u \quad d \quad u$$
$$s \quad u \quad \bar{s} \quad \bar{s}$$
$$d.$$

We see that the s quark in K^- is unaccounted for, which must be gone to make up X. One u quark in the proton is unaccounted for. X must contain this missing u quark, but in which favor? The clue to this lies in the two $\bar{s}$ quarks in K^+ and K^0. This suggests that there must be two s quarks produced after the reaction, and they must be contained in X. Thus the new particle has an sss combinations, and it is the Ω^- particle predicted by Murray Gell-Man (discovered in 1964).

9.3 Spontaneous Symmetry Breaking

The universe was dominated by different particles and interactions during different epochs or phases, and the interaction of the particles are characterized by various symmetries. The present standard model of particle physics exhibits $SU(3)_c \otimes SU(2)_w \otimes U(1)_{B-L}$ symmetry, with $SU(3)_c$ for color symmetry, and $SU(2)_w$ for weak isospin symmetry. It is beyond the scope of this book to discuss these gauge symmetries.

In our present matter-dominated universe most of the gauge symmetries are broken, except for the color symmetry $SU(3)_c$ and the combined discrete symmetry CPT. At each phase transition and symmetry-broken, the physics changed radically. That is what we are interested in and are able to comprehend.

When we say that a system has certain symmetry, we mean that the system doesn't change under some particular transformation. Spherical symmetry, for example, means that a system does not change when we apply a rotation through any angle about any axis through a particular point.

Symmetries have an even a deeper importance in physics. When there is symmetry, there is some quantity that is constant throughout the problem. This means that there is an invariance quantity or a conservation law. For examples, the fact that the laws of physics are not changed by the rotation of a coordinate system leads to the conservation of angular momentum; the fact that the laws of physics are independent of a translation of the coordinate origin leads to conservation of momentum; and the fact that the laws of physics are independent of when we start timing leads to conservation of energy.

We can understand the various forces by understanding what symmetries they have, or, equivalently, what conservation laws they obey. If a reaction that we think will take place does not, it means that there is some conservation law that we might not be aware of, and this reaction violates that conservation law. For example, before the quark theory had been proposed, there was a group of particles that should have decayed by the strong nuclear force, but did not. Because of this strange behavior, these particles were called strange particles. It was proposed that there must be some property of these strange particles that is conserved. The particular decays would then violate this conservation law. When the quark theory was proposed, the strange quark was included to incorporate this property.

Sometimes we find situations that are inherently symmetric, but which somehow lead to an asymmetric result. This is called *spontaneous symmetry breaking*. A spontaneously broken symmetry occurs when a symmetry is not a property of the individual states of the system, even though it is a symmetry of the equations of the system. For example, a marble at the bottom of a special glass bowl can be found in two positions, as shown in Fig. 9.11. When it is balanced at top A, the initial conditions and the equations for the possible motions of the marble are completely symmetric with respect to rotations about the axis of the bowl. As soon as the marble begins to roll, that rotational symmetry is spontaneously broken and the subsequent description of the system has no such symmetry.

A phase transition can spontaneously create symmetry breaking. Water is a good example. Water can exist in three physical states: the gaseous state (vapor), the liquid state (water), and the solid state (ice). So water has three possible phase transitions: gas-liquid, gas-solid, and liquid-solid. The transition from liquid to solid gives the best illustration of spontaneous symmetry breaking. The liquid and solid states of water have two essential differences:

(1) The liquid water is completely isotropic, with no preferred direction. So the properties of liquid water are the same in every direction. In ice the symmetry is broken. Along the lines of the crystal lattice the physical properties differ from those observed in other directions. Thus, whenever there is a transition between the two states, the symmetry is broken according to the physical theory that describes the properties of water.

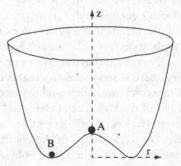

Fig. 9.11 An illustration of spontaneous symmetry breaking.

(2) The crystallized ice has less energy than water in the liquid state. Thus, when we want to melt ice we have to heat it. Conversely, as water turns into ice we have to remove heat energy.

From this familiar example we see that during a phase transition symmetry might be broken spontaneously and energy is either absorbed or released.

As a second example of a spontaneously broken symmetry, consider a magnetized iron bar. An iron bar magnet heated above the Curie temperature 770°C loses its magnetization. The minimum potential energy at that temperature corresponds to a completely random orientation of the magnetic moment vectors of all the electrons, so that there is no net magnetic effect and the bar magnet possesses full rotational symmetry (see Fig. 9.12a). The point of zero magnetization corresponds, using the language of quantum field theory, to the ground or vacuum state; it is the lowest energy state.

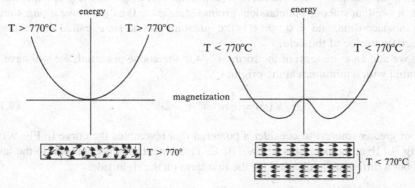

Fig. 9.12 (a) Above the Curie temperature the magnet possesses full rotational symmetry. (b) Below the Curie temperature the full rotational symmetry is spontaneously broken.

As the bar magnet cools below a temperature of 770°C, however, this symmetry is spontaneously broken. When an external magnetic field is applied the electron magnetic moment vectors align themselves, producing a net collective macroscopic magnetization. The corresponding curve of potential energy has two deeper minima symmetrically on each side of zero magnetization (Fig. 9.12b). They distinguish themselves by having the north and south poles reversed. Thus, the vacuum state of the bar magnet is in either one of these minima, not in the state of zero magnetization that is now a false vacuum state. The rotational symmetry has then been replaced by the lesser symmetry of parity, or inversion of the magnetization axis. Note that the potential energy curve in Fig. 9.11b has the shape of a polynomial of at least the fourth degree.

We now take an example from Roos' book as our last example of spontaneous-symmetric breaking; we consider the vacuum filled with a real scalar field $\phi(x)$, where x stands for the space-time coordinates. The potential energy in the vacuum is of the form

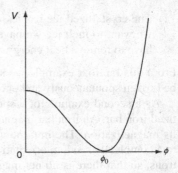

Fig. 9.13 Potential energy of the form (1) of a real scalar field ϕ.

$$V(\phi) = \frac{1}{2}m^2\phi^2$$

which is the familiar parabolic curve with a minimum of $\phi = 0$. If ϕ is a quantum field, it oscillates about the classical ground state $\phi = 0$ as we move along some path in space-time, and $< \phi >= 0$ is the quantum ground state, called the vacuum expectation value of the field.

If we add an extra term of the form $\lambda\phi^4/4$ to the above potential, we still have a potential with a minimum at the origin:

$$V(\phi) = \frac{1}{2}m^2\phi^2 + \frac{1}{4}\lambda\phi^4. \qquad (9.1)$$

It is of greater interest to consider a potential that resembles the curve in Fig. 9.10 or Fig. 9.11b at temperatures below 770°C. This potential is very similar to the last one, but a different sign is used for the first term on the right side:

$$V(\phi) = -\frac{1}{2}\alpha^2\phi^2 + \frac{1}{4}\lambda\phi^4 \qquad (9.2)$$

Its two minima are at

$$\phi_0 = \pm\alpha/\sqrt{\lambda} \qquad (9.3)$$

as shown in Fig. 9.13 Suppose that we are moving along a path in space-time from a region where the potential is given by (9.1) to a region where it is given by (9.2). As the potential changes, the original vacuum $< \phi >= 0$ is replaced by the vacuum expectation value $< \phi_0 >$. Regardless of the value of ϕ at the beginning of the path it will end up oscillating around $< \phi_0 >$ after a time of the order of μ^{-1}. We say that the original symmetry around the unstable *false vacuum* point at $\phi = 0$ has been broken spontaneously.

9.4 Unification of Forces (Interactions)

Experiments indicate that the four fundamental forces act in very different ways from each other at the energies that we are currently able to achieve. However, there is some theoretical evidence that led physicists to believe that at extremely high energies the four forces act in a very similar way. In other words, at extremely high energies the four forces can be seen as a single force of nature. Physicists will not find such high energies in accelerators, but in the early history of the universe particle reactions should have taken place at these high energies. This situation has created an intimate relationship between particle physics and the Big Bang, and has caused many particle physicists to become cosmologists.

Physicists always passionately pursue unification. We can describe the progress of the physical sciences as proceeding from experiments and observations to the discovery of laws, and from a collection of laws to the construction of theories. These are processes of unification. For example, from the planetary motion around the sun and the apple falling on Earth Newton interpreted these two motions in a single unified theory, the universal law of gravity. Electricity, magnetism, and light existed independently before Maxwell; he united them in one theory, the theory of electromagnetism. Einstein spent over 20 years (until his death in 1955) trying to unite electromagnetism with gravity. Although he was never successful, Einstein set the stage for later work that was able to take into account the strong and the weak forces as well.

In the late 1960s Steven Weinberg, Abdus Salam, and Sheldon Glashow successfully united the electromagnetic and weak forces. They showed that, at energies much greater than $100\,\text{GeV}$, the photon, W^+, W^-, and Z^0 all behave in a similar manner. At lower energies the symmetry breaks down, and the weak and electromagnetic force behave quite differently. The complete theory is known as the electroweak theory. For this work they received the Nobel Prize in 1979.

Merging the electromagnetic and weak forces was an ambitious project. The two forces seem quite distinct from each other. The electromagnetic force is effectively infinite in range, whereas the weak force is confined to distances less than 10^{-18} m. The electromagnetic force acts only between charged objects, but the weak force can act between neutral objects. Furthermore the weak force changes the nature of the particles, and the weak force needs three massive exchange particles (W^+, W^-, Z^0) to mediate, while the electromagnetic force only needs one massless exchange particle, the photon.

It is beyond our scope to reproduce the mathematical equations of electroweak theory, but the basic idea goes this way. Salam, Weinberg, and Glashow started with weak and electromagnetic fields and four exchange particles—W^+, W^-, Z^0 and the photon γ, all no mass. In addition they introduced the Higgs field, which only interacts with the weak field but not with the electromagnetic field. As the W bosons pass through space they interact with the Higgs field, an exchange of energy takes place, and the result of this is that the W bosons take on mass. The W^0 boson thus modified is able to interact with a photon and turn into a new boson Z^0, which has a mass different from W^+ and W^-. Furthermore, the Z^o interacts with other particles in a very specific way—it does not change the type of particle (in that way it is more like a photon than a W), and it can couple to objects with no charge (unlike the photon). This produces an effect known as "neutral current," in which neutrinos can be seen to interact with other particles and stay as neutrinos. This type of reaction was discovered in 1973 and provided an early pointer to the theorists that they were heading in the right direction.

The electroweak theory was confirmed in 1983 at CERN with the discovery of the W^+, W^-, and Z^o bosons, their masses tied exactly with the theory.

The success of the electroweak theory led to a number of attempts that try to combine the electroweak theory with the strong force into what is called a grand unified theory (or GUT). This title is rather an exaggeration: the theories are not all that grand, nor are they fully unified, as they do not include gravity. Nor are they really complete theories, because they contain a number of parameters whose values cannot be predicted from the theory but have to be chosen to fit in with the experiment. Nevertheless, they may be a step toward a complete, fully unified theory.

Although in specific details one GUT may differ from another, the basic idea is as follows: The strong force gets weaker at high energies, and the weak and electromagnetic forces get stronger at high energies. At some very high energy, called the grand unification energy, these three forces would all have the same strength and can be seen as the disturbances of a set of basic fields linked via a Higgs field. As a consequence there should be new fields that have not yet been seen with their own disturbances, the X and Y particles. The detailed properties of these particles depend on the exact version of the theory used, but the masses of these particles must be about 10^{14} GeV/c^2 (Fig. 9.14). We can see the general idea behind this by

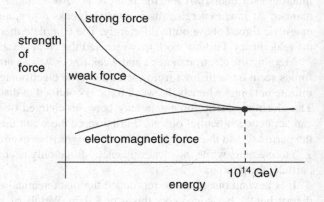

Fig. 9.14 The variation of force with energy.

combining Table 9.7 for the u quarks and Table 9.8 for the electroweak interaction. The result is Table 9.9, with five rows and five columns, which is partially filled:

Table 9.9 Example of Speculative GUT Particles

	e	ν	uR	uG	uB
e	γ; Z^0	W^-		X	
ν	W^+	Z^0			
uR			g(R-B)	g(R-G)	g(R-B)
uG	X	g(G-R)	g(G-G)	g(G-B)	
uB			g(B-R)	g(B-G)	g(b-b)

The top left section is filled with intermediate bosons (W^+, W^-, and Z^o) and the photon. At bottom right we only show the gluons associated with the strong interaction. To these we should add the intermediate bosons and photons that are also coupled to quarks. Two empty sections are marked X. The principle of GUT supposes that there are new bosons to fill the gaps. These new particles (12 is the number that emerges in the simplest case) carry an electric charge, a weak charge, and color, and they allow interactions between leptons and quarks.

The electric charge of these hypothetical X bosons has fractional value ($-4e/3$, $e/3$, $+e/3$, or $4e/3$), and they are also extraordinarily massive, some 10^{15} times the mass of the proton. With this huge mass, the distance over which they exert an appreciable force is an extremely small 10^{-29} cm (given by the Compton wavelength). Finding such tiny yet extraordinarily massive X bosons is out of reach of present techniques. Many particle physicists have turned their attention to a hot Big Bang, where the available energy in the earliest epoch was large enough to create X particles. These should have left some observable traces, such as evidence for the inflationary phase of the universe, which will be discussed in the next chapter.

Although we do not know which GUT is correct at the moment, they all predict that protons are not stable particles. The quarks inside should be able to emit X and Y particles and turn into leptons. The lifetime of the proton is predicted to be more than 10^{30} years! But this doesn't stop experimental particle physicists from trying to detect it.

The value of the grand unification energy is not very well known; it would probably have to be 10^{14} to 10^{15} GeV. This is well beyond the range of any conceivable particle accelerator. The greatest hope that the theorists have of testing their theories lies in speculating about the early universe. Although the three interactions all have the appearances of being symmetrical and unified, they do not manifest themselves on an equal basis under all circumstances. This is due to the large differences between the masses of the vector particles—they are 0, 100, and 10^{15} times the mass of the proton. If particles were colliding at energies greater than 10^{14} GeV (equivalent to a temperature of 10^{27} K, $E \sim kT$), the strong, weak, and electromagnetic interactions would be indistinguishable from each other. As the energy of the colliding particles fell below 10^{14} GeV, the X bosons could no longer be created spontaneous from the ambient energy, and the strong nuclear force is no longer unified with the electromagnetic and weak interactions. That is, at energy 10^{14} GeV (or at temperature 10^{27} K), there was a spontaneous breaking of symmetry. This situation existed in the very early universe. The critical temperature 10^{27} K corresponds to 10^{-35} second. As the universe expanded further, at temperature 10^{15} K intermediate bosons were toppled also. This critical temperature corresponds to 10^{-11} second. The spontaneous breaking of symmetry at the two critical temperatures plays a crucially important role in releasing the latent energy of the vacuum as implied by the Higgs particles.

Most GUTs require that neutrinos have mass, approximately given by

$$m_\nu \approx M_{eW}{}^2 / M_x$$

where M_{eW} is a characteristic mass of the electroweak interaction, roughly $10^2 \, \mathrm{GeV}/c^2$, and M_x is the unification mass $E_x/c^2 \approx 10^{15} \, \mathrm{GeV}/c^2$. Nearly all GUTs project M_x values of this order of magnitude, which in turn means that m_ν is less than $1 \, \mathrm{eV}$. The theory also predicts $m(\nu_e) \ll m(\nu_\mu) \ll m(\nu_\tau)$, that all of the neutrino masses are inaccessible to direct measurement with existing technology. Even so, the impact of massive neutrinos on both the solar neutrino problem and the cosmological "dark matter" is substantial.

9.5 The Negative Vacuum Pressure

According to current quantum thinking, all of the particles are manifestations of various fields. Quantum theorists believe that space is not empty, but filled with these fields. Their energy fills the vacuum space. Vacuum is not nothing; it is the state of lowest possible energy, and fluctuations of this energy can cause particles to appear. So the vacuum is not simple; it is an active place, a sea of continuously appearing and disappearing particles (thanks to Heisenberg's uncertainty principle). Some physicists believe that fields are everywhere in the universe and that they are the simplest irreducible fundamental entities of physics. Perhaps all the particles of nature are fully defined by field equations that describe their properties and their interactions. According to S. Weinberg, all of reality is a set of fields. Everything else can be derived from the dynamics of quantum fields.

A very complicated problem is that of the energy density of a vacuum. It turns out that the energy of the vacuum always enters the formulae of the theory in such a way that it is ultimately canceled out when the formulae are applied to real particle systems. The theory may be reformulated in such a manner that the average energy density of the vacuum becomes exactly zero. This approach, however, is justified only as long as the gravitational interaction of virtual particles is not taken into account.

In the late 1960s the Soviet physicist Zel'dovich put forward arguments that show in simple terms how a nonzero energy density of the vacuum could emerge. Virtual particles with rest mass m (for simplicity we consider one kind of particle) are being created and annihilated in the vacuum. The average density of proper mass (or of proper energy, the quantity differing from the mass density according to $E = mc^2$ by the factor c^2 only) of virtual particles does not enter the final expressions and may be set equal to zero, as mentioned above. Quantum theory associates a characteristic length $l = \hbar/mc$ with any particle of mass m, where $\hbar$ is Planck's constant divided by 2π. The average distance of separation of a newly born pair of virtual particles is about the characteristic length l. The energy of the gravitational interaction of such a pair can be estimated from the conventional formula:

$$E = Gm^2/l.$$

It is this energy that can give rise to the nonzero energy density of the vacuum, or, correspondingly, to a nonvanishing mass density of the vacuum $\rho_{vac} = \varepsilon_{vac}/c^2$.

To estimate the density of energy ε_{vac}, we divide E by the volume l^3 occupied by one virtual particle

$$\varepsilon_{vac} = (Gm^2/l)/l^3 = Gm^6c^4/\hbar^4.$$

The last term in the above equation is obtained by the substitution of $\hbar/mc$ for l.

As energy density is just the energy per unit volume, when the volume of space increases, the total vacuum energy increases correspondingly. In other words, the larger the volume, the greater the total energy.

This is not all, however. The theory requires also that the "vacuum fluid" exert some pressure, but, in contrast to pressure in the usual sense, this vacuum pressure must be negative.

What do we mean by negative pressure? Is that not contradictory? For ordinary systems, a positive pressure results as the energy increases upon compressing; the increase in pressure resists further compression. This is a familiar phenomenon: push down the piston of a cylinder of gas, the gas pressure goes up; if the piston is pulled outward, the gas pressure goes down, and the energy density also goes down. Negative pressure behaves oppositely. Although the concept of a negative pressure seems strange at first, it really means only that the associated total energy increases when the volume of the system increases, rather than decreasing with increasing volume as does ordinary positive pressure. We can demonstrate this mathematically, with a formula from Einstein's theory. The absolute magnitude of the vacuum pressure must be equal to that of the energy density, i.e., $p_{vac} = -\varepsilon_{vac}$; making use of the Einstein relationship, this becomes $p = -\rho c^2$, or $p/c^2 = -\rho$, where for simplicity we have dropped the subscript vac. Now recall that the mass of a uniform sphere of radius R is $M = (4\pi/3)R^3\rho$. This formula is valid when $p < \rho c^2$. Otherwise we have to use the following formula

$$M = \frac{4}{3}\pi R^3(\rho + 3p/c^2).$$

Under the usual circumstance $p \gg \rho c^2$, so the term $3p/c^2$ can be neglected. This is not the case for virtual particles in a vacuum, for which the pressure and the energy density of gravitational interaction are linked through the relation $\rho_{vac} = -p_{vac}/c^2$.

Assuming the "vacuum fluid" uniformly fills the whole space, we can calculate the gravitational acceleration caused by such a fluid easily:

$$a = -\frac{GM}{R^2} = -\frac{4}{3}\pi G(\rho_{vac} + 3p_{vac}/c^2)R$$
$$= -\frac{4}{3}\pi G(\rho_{vac} - 3\rho_{vac})R = \frac{8}{3}\pi G\rho_{vac}R.$$

This result shows that the gravity of a vacuum is not attractive, as for ordinary matter, but repulsive. Note that the sign of the value of a in the above equation is positive! Such a repulsion apparently stems from the fact that the vacuum pressure is negative and participates in the gravitational interaction, as Einstein's theory shows, on a par with the energy density.

In the nonintuitive world governed by general relativity a positive pressure tends to make the universe collapse. Thus we expect that a negative pressure would cause it to expand and, at least in the very early stage, expand faster and faster.

References

Allday J (1998) *Quarks, Leptons and the Big Bang*. (Institute of Physics Publishing, Bristol, England)

Carrigan RA, Trower, WP, eds. (1989) *Particle Physics in the Cosmos*. (WH Freeman and Co., New York)

Roos M (1994) *Introduction to Cosmology*. (John Wiley & Sons, New York)

Chapter 10
The Inflationary Universe

We have so far glossed over the drawbacks of the standard Big Bang theory. We now address two of the drawbacks in this chapter: the flatness problem and the horizon problem. In the early 1980s, Alan Guth resolved these two problems with his inflationary theory. His basic idea is that the universe enters a false vacuum state shortly after the Big Bang, then tunnels out and expands exponentially. We choose to discuss Guth's original model (now called classical model or old inflation) for pedagogic reasons. Guth's model has many nice qualitative features; it does not work quantitatively. Therefore, A. Linde, A. Albrecht, P. Steinhardt, and others constructed new models as remedies. It is not clear which of the new models is correct, so we will discuss each of them briefly.

10.1 The Flatness Problem

In Chapter 9 we discussed how the density of matter of the universe would determine its future. If the density parameter Ω (ratio of the density divided by the critical density) is less than unity, the expansion of the universe will continue forever; if Ω is greater than unity, the resulting gravity will be strong enough to eventually halt the expansion of the universe; and if $\Omega \cong 1$, then the universe is marginally bound, and space is almost flat.

The average density of the luminous matter is about 10% to 20% of the critical density. But, evidence for the existence of significant amounts of dark matter suggests that the true density of matter (luminous and dark matter) across the universe may be equal to the critical density. This means that the universe is marginally bound and space is flat. Was Ω always approximately equal to unity? Let us take a close look at this.

As there is a constant amount of matter in an expanding universe, so the density of the universe changes with time. It is not as obvious that the critical density also changes with time. The critical density is the amount of matter per unit volume required to provide enough gravitational pull to halt the expansion of the universe.

As the force of gravity depends on distance, its pull on parts of the universe farther away decreases as the space gets bigger. Thus, the amount of matter required to close the universe is different from what it would have been in the past. In other words, the critical density changes with the age of the universe.

Therefore, both the density of the universe and the critical density change with the age of the universe, and the density parameter Ω is not constant. As Ω evolves it will always stay ≥ 1 if it started that way; and it will always be ≤ 1 if that was the value it started with. The fate of the universe was determined at the moment of the Big Bang. So what was the value of Ω initially?

To answer this question, let us recall equation (9-42) that we now rewrite as

$$\frac{\dot{R}^2}{R^2} + \frac{kc^2}{R^2} = \frac{8\pi G \rho_o}{3}$$

where $R(t)$ is the scale factor, $\dot{R}$ its rate of change with time, ρ_o the density (matter and energy) of our universe, and the curvature parameter k can take the values $1, 0, -1$. In terms of the density parameter Ω:

$$\Omega = \frac{\rho_o}{\rho_c} = \frac{8\pi G \rho_o}{3H^2}$$

the Friedmann dynamical equation (9-40) becomes

$$\Omega = 1 + \frac{kc^2}{H^2 R^2} = 1 + \frac{kc^2}{\dot{R}^2}.$$

If $k = 0$, we have $\Omega = 1$. For $k \neq 0$, as we approach closer and closer to the Big Bang epoch, $\dot{R}$ increases, and the second term on the right-hand side of the above equation becomes smaller and smaller; accordingly, the density parameter approaches 1. In other words, at the beginning, the density of the universe must have been at or near the critical density point.

The problem is that given all of the infinite possible masses that our universe could have, why does it have a mass so close to the critical value? Because $\Omega = 1$ means that our universe is flat, so this puzzle is called the flatness problem.

The earliest understandable moment in the universe was the Planck time, which is about 10^{-43} s after the Big Bang. Before the Planck time, the universe was so dense and particles were interacting so violently that no known theory can describe what was happening then. What, then, could have happened immediately after the Planck time to ensure that $\Omega = 1$ to a very high degree of accuracy?

10.2 The Horizon Problem

A second question, closely related to the flatness problem, is called the horizon problem. It concerns the isotropy of the cosmic microwave background radiation. The cosmic background radiation deviates from complete uniformity by only 1 part in 100,000! This tells us that the universe was extremely isotropic at the time of decoupling, and the background radiation was emitted from regions with a common temperature. All parts of the universe must have been interacted with each other before that time. How did this happen? Therein lies the problem.

W. Rindler pointed out the horizon problem in 1956. He introduced a quantity called the horizon distance D to define and explain the horizon problem. D is the age of the universe (t_0) multiplied by the velocity of light (c):

$$D = c \times t_0.$$

No information can be received from a distance farther than D, which is called the horizon distance (or cosmic distance). The visible universe is defined by the horizon distance D, which is about 13 billion light-years, or 3×10^{27} cm (Fig. 10.1). As the universe expands, D increases faster than the radius R of the universe. If we go back in time, D decreases faster than R.

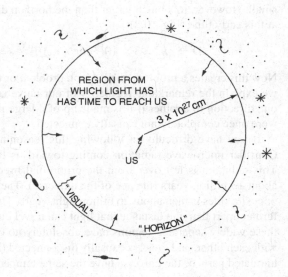

REGION FROM WHICH LIGHT HAS HAS TIME TO REACH US

3×10^{27} cm

US

"VISUAL"

"HORIZON"

Fig. 10.1 The visible universe is defined by the horizon distance.

Now let us retrace back to our visible universe. The material it contains today was contained within a much smaller region in the past. As the universe expands, the mean photon energy decreases as R^{-1} because of the redshift, so that the temperature T, which is proportional to this mean energy, also decreases as R^{-1}. This means that we can use the radiation temperature as a gauge of the size of the currently visible universe in the past.

Let us go back to 10^{-35} s after the Big Bang; this is the epoch of grand unification and the temperature the universe was about 3×10^{28} K. Today, after about 10^{17} s(13×10^9 years) of expansion, the temperature of radiation has fallen to about 3 K. So, the temperature has changed by a factor of 10^{28} since that early time. The contents of the universe as we see it today were contained in a sphere 10^{28} times

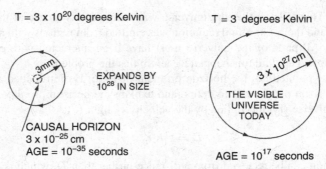

T = 3 x 10^{20} degrees Kelvin

T = 3 degrees Kelvin

3mm

EXPANDS BY
10^{28} IN SIZE

3 x 10^{27} cm

THE VISIBLE
UNIVERSE
TODAY

CAUSAL HORIZON
3 x 10^{-25} cm
AGE = 10^{-35} seconds

AGE = 10^{17} seconds

Fig. 10.2 Retrace the history of our visible universe to 10^{-35} s.

smaller than now. This is equal to $(3 \times 10^{27}\,\text{cm})/10^{28}$, or 3 mm. This is amazingly small. However, it is much larger than the horizon distance of about 3×10^{-25} mm at this early time (Fig. 10.2):

$$D = ct_0 = 3 \times 10^{10}\,\text{cm/s} \times 10^{-35}\,\text{s} = 3 \times 10^{-25}\,\text{mm}.$$

Now this creates a problem–the horizon problem or the isotropic problem. How can we explain the remarkable regularity of our universe from place to place and from one direction to another if it is made up of a large number of separate regions that were once completely not causally connected?

If you have difficulty in following this reasoning, try the following approach. Consider microwave radiation coming toward us from opposite sides of the sky. This radiation is left over from the primordial fireball and has been traveling for about 13 billion years (the age of the universe). The total distance between the two opposite sides is then about 26 billion light-years. That is, the two opposite sides are farther apart than the distance that light can travel during the age of the universe, so these widely separated regions have absolutely no connection (or communication) with each other, and they are causally disconnected (Fig. 10.3). How, then, can these unrelated parts of the universe have the same temperature?

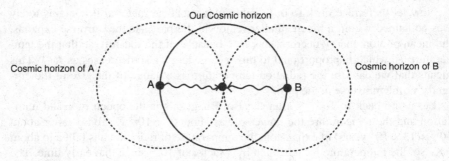

Our Cosmic horizon

Cosmic horizon of A

Cosmic horizon of B

A

B

Fig. 10.3 The horizon problem.

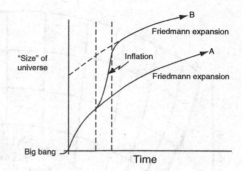

Fig. 10.4 The observable universe with and without inflation.

10.3 Alan Guth's Inflationary Theory

In the early 1980s, Alan Guth found a way of getting around the horizon and flatness problems. He proposed a model called inflation (Fig. 10.4): the early universe experienced a rapid (exponential) expansion, between 10^{-35} and 10^{-32} seconds due to a phase change, leading to the introduction of energy into the universe with the effect of antigravity. Guth discovered this while examining how the fundamental forces could be unified into a single force. At high energies in the early universe, it is thought that the forces were unified (GUTs), but as the universe cooled, the strong force became distinct from the others. This had the effect of a phase change. During the inflationary expansion period, the universe might have increased in size by a factor of 10^{50} or more, from a region of space smaller than a proton to a volume about the size of a grapefruit, at which point the Hubble expansion resumes again.

The key ingredient of Guth's inflationary scenario is the assumed occurrence of a phase transition in the very early universe; this phase transition is also linked to spontaneous symmetry breaking. Grand unified theories imply that such a phase transition occurred in time before 10^{-35} sec and at a temperature above 10^{27} K. At temperatures higher than 10^{27} K there is one unified type of interaction, while at temperatures below 10^{27} K, the strong force broke off and the grand unified symmetry was broken.

When the universe cooled down to the temperature of this phase transition, either of two things may have happened: the phase transition may have occurred immediately, or it may have been delayed, occurring only after a large amount of supercooling. The word supercooling refers to a situation in which a substance is cooled below the normal temperature of a phase transition without the phase transition taking place. For example, steam can be supercooled below the boiling point of water. The supercooled state has a higher energy and so it is unstable; with a slight disturbance it condenses to a bubble of water. Whether the infant universe behaves similarly and nucleates into bubbles depends on the physics of GUTs. If the correct GUT and the values of its parameters were known, there would be no ambiguity about the nature of the phase transition; we would be able to calculate how quickly it would occur. In the absence of this knowledge, however, either of the two possibilities appears plausible. But calculations show that only an extremely narrow range of parameters

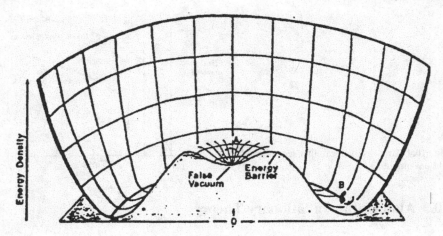

Fig. 10.5 The universe is trapped in the false vacuum state.

leads to an intermediate situation; in almost all cases the phase transition is either immediate or strongly delayed. So Guth assumed that the GUT phase transition and symmetry breaking did not take place immediately. As the universe supercooled to temperatures far below the temperature of the phase transition, it would have approached a very peculiar state of matter called a false vacuum (Fig. 10.5). At such small scales and high energies, quantum physics supercedes classical physics, and it allows phase transition to take place by means of "quantum tunneling."

The false vacuum has a peculiar property that makes it very different from any ordinary material. For ordinary material the energy density is dominated by the rest energy of the particles of which the material is composed ($E = mc^2$). If the volume of an ordinary material is increased, the density of particles will decrease, and therefore the energy density also decreases. The false vacuum, on the other hand, is the state of lowest possible energy that can be attained while remaining in the phase for which the grand unified symmetry is unbroken. Its energy is attributed to the Higgs fields, which are included in the theory to produce a unified theory and spontaneous symmetry breaking. Remember that we are assuming that the GUT phase transition occurs very slowly, so for a long time (by the standard of the very early universe) the false vacuum is the state with the lowest possible energy density that can be attained. Thus, even as the universe expands, the energy density of the false vacuum remains constant.

How can we hold the energy density fixed while space is in the process of expanding? If the new space being added also contributes energy as it comes into being, then the total energy density of the false vacuum can indeed remain relatively constant.

We have learned in the preceding chapter that the false vacuum behaves like a gas with negative pressure. When this peculiar property of the false vacuum is combined with general relativity, we get a dramatic result: the false vacuum provides a gravitational repulsion. So when the universe was caught in the false vacuum state, gravity caused the expansion to accelerate. Some claim that the form of this repulsion is identical to the effect of Einstein's cosmological constant. But the repulsion caused by a false vacuum operates for only a very limited period of time.

Alan Guth showed that this cosmic repulsion has the effect of stimulating the expansion of the universe at an explosive exponential rate. The result is that space triples its size every 10^{-34} second after the epoch 10^{-35} s. This exponential expansion is termed inflation and would last until the epoch 10^{-32} s, as the theory suggests. To show that the expansion is indeed exponential, let us go back to Friedmann's equations, which take the following forms in the present context:

$$\frac{\dot{R} + kc^2}{R^2} = \frac{8\pi G}{3c^2}(u_r + u_v)$$

and

$$2\frac{\ddot{R}}{R} + \frac{\dot{R}^2 + kc^2}{R^2} = -\frac{8\pi G}{c^2}(p_r + p_v)$$

where u_r and p_r are the energy density and pressure of the relativistic particles, with $u_r = 3p_r > 0$ and u_v and p_v are the energy density and pressure of the false vacuum, with $p_v = -u_v < 0$. The quantities u_r and p_r decrease with the expansion of the universe as $1/R^4$, but p_v and u_v stay constant as long as the false vacuum is maintained. Thus u_v and p_v tend to dominate the behavior of the solution to the equations, and also dominate the curvature term, k/R^2. Under these conditions, the last two equations take the simple forms

$$\frac{\dot{R}^2}{R^2} = a^2 \quad \text{and} \quad 2\frac{\ddot{R}}{R} + \frac{\dot{R}^2}{R^2} = 3a^2$$

from which we have

$$\ddot{R} = a^2 R,$$

and

$$R(t) \approx R(0)\exp\left(a^2 t\right)$$

i.e., R has the exponential behavior described earlier.

As there are 100 units of 10^{-34} second to use up before the elapsed time is 10^{-32} second, the tripling process takes place around a hundred times: $3 \times 3 \times 3 \dots \dots = 10^{50}$ times altogether. During this time the universe would inflate to about the size of a basketball.

Once the universe tunnels through the energy barrier and into the true vacuum state (Fig. 10.6), the rapid exponential inflation stops. The GUT phase transition is completed, and the latent energy is released, resulting in a tremendous amount of particle production. The universe is reheated in the process to almost 10^{27} K. From this point on, the expansion of the universe slows to the regular pace of the Hubble expansion.

As any particle density present before inflation would have been diluted to a negligible value by the enormous expansion, in the inflationary theory virtually all the matter and energy in the universe were produced during the inflationary

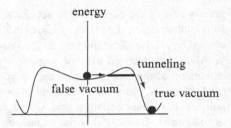

Fig. 10.6 The universe tunnels through the energy barrier from the false vacuum.

process. This seems strange: how could it be possible that all the energy in the universe was produced as the system evolved? Is this violating the principle of energy conservation?

The loophole in the conservation of energy argument is due to the peculiar nature of gravitational energy. Energy has the capacity to be either positive or negative. Two objects attracted by the force of gravity need energy to pull them apart, and therefore in that state we say that they have negative gravitational energy. In other words, we can say that negative energy is stored in the gravitational field. According to Newtonian theory, the gravitational energy of a mass m attracted by another mass M is given by,

$$E_g = -GmM/R,$$

where R is the distance between the two masses, and G is the gravitational constant. Two objects that are close to each other have less energy than the same two objects farther apart, which means that energy can be extracted as the two masses come closer. Once the two masses come together, their gravitational fields will be superimposed, producing a much stronger gravitational field. The net result of such a process is the extraction of energy and the production of a stronger gravitational field.

To apply the last equation to our universe, M denotes the net mass of the universe contained within the Hubble radius $R = c/H$, where H is Hubble's constant. For a universe that is approximately uniform in space, this negative gravitational energy may exactly cancel the positive energy represented by the matter, so the total energy of the universe is zero. Edward Tyron speculated this as early as 1973. He noticed that observations seem to indicate that our universe is probably closed, in which case the mass density of the universe exceeds the critical density ρ_c ($\rho_c = 3H^2/8\pi G$). Using the critical density in our estimate of E_g, we obtain:

$$E_g = -\frac{GmM}{R} = -\frac{Gm}{R}\left(\frac{4\pi}{3}R^3\rho_c\right) = -\frac{mc^2}{2}.$$

Hence, within a factor of 1/2, the negative gravitational energy of any piece of matter is sufficient to cancel the positive mass energy of mc^2. This simple argument indicates that the net energy of our universe may indeed be zero.

P. G. Bergmann has presented a more sophisticated argument that indicates that any closed universe has zero energy. In its simplest form, the argument is as follows:

Suppose the universe were closed. Then it would be impossible for any gravitational flux lines to escape. If a viewer in some larger space in which the universe were embedded were to view the universe, the absence of gravitational flux would imply that the system had zero energy. Hence a closed universe has zero energy.

If the net energy of our universe is indeed zero, then it could have come into existence from nowhere as a result of quantum fluctuation of the vacuum. Fluctuations are very familiar to physicists: particle-pairs are created and annihilated constantly in emptiness, for a period allowed by the uncertainty principle. However, it is still very speculative to suggest that a very large universe may have appeared as a fluctuation of the vacuum and then survived for a very long time. We shall not pursue this question further here and instead will return to Guth's inflationary scenario.

As the universe expanded, the energy of the gravitational field became more and more negative, but the energy stored in the false vacuum became larger and larger. As an analogy, this may be compared with a block of rubber; the more the rubber is stretched, the more energy it has because the elastic fibers store energy.

How do we know whether inflation happened? The microwave background radiation may provide the test. In the late 1990s COBE observations of the microwave background showed small structures consistent with inflation. More observations need to be done. In summer 2001, NASA launched the Microwave Anisotropy Probe, and in 2007 the European Space Agency's *Planck* spacecraft will conduct detailed mapping across the entire sky.

10.4 The Successes of Guth's Inflationary Theory

10.4.1 The Horizon Problem Resolved

Guth's inflationary universe theory appeared to solve the horizon and flatness problems we have discussed. Let us see how the theory accounts for the isotropy of the microwave background. As depicted in Fig. 10.7a, in a noninflationary model,

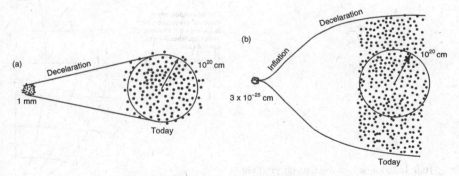

Fig. 10.7 Inflation solves the horizon problem.

today's observable universe would have expanded from a region of 3 mm across at 10^{-35} sec after the Big Bang. Even though this is very small, it is still much larger than the horizon distance of the universe at that time. However, in an inflationary universe model, space was expanded, from a much smaller region of about 3×10^{-25} cm, during the period of inflation to become much larger than the horizon distance of the universe (Fig. 10.7b). Thus, in examining microwaves that are from opposite sides of the sky, what we are seeing is radiation from parts of the universe that were originally this phase transition linked to the spontaneous symmetry breaking in intimate contact with each other. This common origin is why they have the same temperature.

10.4.2 The Flatness Problem Resolved

The flatness problem can also be solved by inflation. Recall that

$$\Omega = 1 + \frac{kc^2}{H^2 R^2} = 1 + \frac{kc^2}{\dot{R}^2}.$$

In the standard Big Bang theory, the expansion of the universe is slowing down, so the second term on the right-hand side increases with time, forcing Ω away from one. Inflation reverses this state of affairs, because

$$\ddot{R} > 0 \Rightarrow \frac{\mathrm{d}}{\mathrm{dt}}(\dot{R}) > 0 \Rightarrow \frac{d}{dt}(RH) > 0.$$

So, the condition for inflation is precisely that which drives Ω towards one rather than away from one.

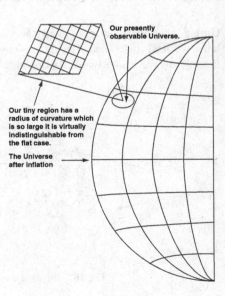

Fig. 10.8 Inflation solves the flatness problem. (Adopted from Guth 1997.)

Figure 10.8 might help us to visualize how inflation theory solved the flatness problem. Any curvature to space-time is stretched out by inflation, like a cosmic version of a balloon stretched to enormous proportions. After inflation, our universe is a tiny region on the surface of a much larger curved surface. We can think about a small portion of Earth's surface, such as our backyard. For all practical purposes, it is almost impossible to detect Earth's curvature over such a small area, and so our backyard looks flat. Similarly, the observable universe is such a tiny fraction of the entire inflated universe that any overall curvature in it is undetectable.

Note that Guth's inflationary scenario doesn't violate Einstein's dictum that nothing can travel faster than the speed of light, because the expansion of the universe is the expansion of space and does not involve the motion of matter or energy through space.

10.5 Problems with Guth's Theory and the New Inflationary Theory

The first reactions to Guth's theory were very favorable. However, when the theory was investigated further, problems began to appear. To understand these problems, let us revisit the problem of how the strong force is "frozen" out of GUTs. Anyone who has ever watched a lake freeze over in winter should notice that the ice does not form a film on a lake all at once, but begins in several spots simultaneously, spreading out from these spots until it has covered the entire surface. An ice sheet is formed as follows. The axes of symmetry of the crystals growing around one center will, in all probability, point in a different direction from the axes of symmetry associated with the neighboring center, as shown in Fig. 10.9a. When the ice spreading outward from the two centers comes together, there will be a region where the axis of symmetry changes from one direction to the other. The result is a pattern as shown on the right in Fig. 10.9b, where regions with different directions of symmetry come together. Each of the separate regions is called a domain or a bubble, and the regions between the domains are called domain walls. Guth thought that the strong force "freezing" out of GUTs might proceed in much the same way, with the new level of reduced symmetry being established first over small volumes in space and then growing to fill everything. Thus, the theory predicted that inflationary expansions should have taken place in a lot of separate domains, or spatial

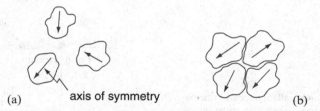

(a)   axis of symmetry   (b)

Fig. 10.9 (a) Ice film forms simultaneously in different spots in a winter lake. (b) A domain or bubble is formed.

bubbles. As the domains expanded, they collided with one another and coalesced into one big universe. But this expectation was not realized; the bubbles tended to cluster in groups around the largest in a group, with each group lying well away from its neighbors. This created a big problem. Let us explain briefly what the problem is. Theory showed that the energy released in phase transition tends to reside on the surfaces of the bubbles, with the largest energy being on the walls of the largest bubble of the typical group. For "reheating" to take place it was essential for the bubble walls to collide. Collisions occurring frequently would release and redistribute the energy residing in individual units; without such a redistribution, reheating of the inflated universe could not be achieved. As a consequence, the universe today would not be extremely homogeneous or isotropic.

Although Guth's original model encountered serious problems, its central attractive feature of supercooling during the phase transition and inflation due to the negative pressures of the false vacuum prompted modifications of the old model, rather than abandonment, by many researchers, including A. Linde in Moscow, A. Albrecht and P. J. Steinhardt at the University of Pennsylvania, and Guth himself. The new versions differ from Guth's old theory in two important ways: (1) although there is a false vacuum, it plays a much less crucial role in the new theories; and (2) the entire observable universe is contained in one bubble only.

In Guth's old model the false vacuum lay in a high valley surrounded by a barrier of mountains that descended to the much lower level of the true vacuum on the other side. So quantum tunneling was needed to penetrate the barrier. Once tunneling took place, bubbles formed. But the growth of a typical bubble after formation was not rapid enough to guarantee collisions between bubbles. In the new theories the false vacuum lies in a shallower valley on a tall plateau, which slopes down very gently for a while, then descends very rapidly down to the true vacuum lying in the valley, as shown in Fig. 10.10.

The phase transition now works differently from Guth's old model. Modest tunneling would let the universe get out of the shallow valley and start a slow roll. As the gentle roll at the high plateau is going on, the conditions of the false vacuum operate within the region, and the region grows exponentially. In the meantime, the phase transition proceeds very slowly, with the Higgs fields growing slightly from their initial zero value. When the rapid descent to the true vacuum occurs, the Higgs fields grow fast. As the Higgs fields reach their final values, the inflation would stop, and Friedmann expansion would take over. The slowness of phase transition now allows inflation go on for much longer time than in Guth's old version, and

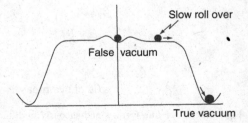

Fig. 10.10 In a new inflation scenario the potential energy has a shallower valley at the top. (Adopted from Narlikar 1988.)

we have a single large region that would vastly exceed the size of the observable universe today.

The universe is reheated by the energy release through rapid oscillations and subsequent decay of the Higgs fields/particles. The unstable heavy Higgs bosons decay profusion into lighter particles; through rapid collisions these lighter particles would quickly achieve thermodynamic equilibrium.

This new model solves the horizon and flatness problems. It also has no exit problem, as well as fewer defects of all kinds, including magnetic monopoles. Finally, it does well in explaining the spectrum of inhomogeneities, which is related to inherent inhomogeneities of the Higgs field prior to inflation. Quantum fluctuations do not allow a completely smooth Higgs field, even in the vacuum state of zero value. These inhomogeneities are on a very small scale prior to inflation, but they grow in size by inflation. They are transferred from the Higgs field to matter after the Higgs bosons decay.

The inflationary model is not a detailed theory, but is just an outline for a theory. To fill in the details, we need to know much more about the details of particle physics at the energy scales of GUTs, and perhaps beyond. The field of particle physics and cosmology will be closely linked for years to come.

10.6 Problems

10.1. During standard Big Bang evolution, the density parameter Ω moves away from one unless its initial value was precisely one. Can Ω become infinite and if so what does this mean?

10.2. The classical example of inflationary expansion is a universe possessing a cosmological constant λ. In this case, the Friedann equation becomes (in terms of H)

$$H^2 = \frac{8\pi G}{3}\rho - \frac{k}{R^2} + \frac{\lambda}{3}$$

where $H = \dot{R}/R$. Show that when the universe is dominated by a cosmological constant, the expansion rate of the universe is given by an exponential function:

$$R(t) = \exp\left(\sqrt{\lambda/3}t\right).$$

(Note that here λ has units $[\text{time}]^{-2}$. Most people measure it as $[\text{length}]^2$. The difference is an explicit factor c^2.)

References

Guth AH (1997) *The Inflationary Universe.* (Addison Wesley, Reading, MA)
Narlikar JV (1988) *The Primeval Universe.* (Oxford University Press, Oxford, UK)
Tryon EP (1973) Is the universe a vacuum fluctuation? Nature 246: Dec.14

Chapter 11
The Physics of the Very Early Universe

11.1 Introduction

All the Friedmann models that were described in Chapter 8 have the common feature that R = 0 at a certain epoch, which we have chosen to label by $t = 0$, and all times are measured with respect to that instant. As we approach $R = 0$, the Hubble constant increases rapidly, becoming infinite at $R = 0$. This epoch therefore indicates violent activity and is given the name Big Bang. At this epoch the mathematical description of space-time geometry breaks down. $R = 0$ is also an insurmountable barrier to physicists: laws of physics break down there also. This doesn't mean that physics will never be able to explain the basic problem in cosmology. But we should move forward with caution and avoid overconfidence in simple extrapolations based on a large body of data from observations and theoretical and experimental advances in particle physics. We are able to describe the development of the universe starting about 10^{-11} seconds after the Big Bang and to follow the behavior predicted by the standard model of particle physics and general relativity.

In this chapter we will concentrate on the physics of the very early universe. Compared with the great age of the universe, nearly everything that is interesting in cosmology occurred at this very early time. The very early universe here means from the Planck time to the radiation-dominated era. To others, very early universe may mean the epoch of the first three minutes after the birth of the universe. Many of the significant events of the early Big Bang took place extremely rapidly, in a tiny fraction of second. Only a few minutes later the universe hardly changed at all during any 1-second interval.

Why do we only try to give account of the evolution of the early universe from Planck time? We actually don't know anything about the universe before this time. Let us see why. Near a mass m, general relativity prevents our seeing events occurring at dimensions less than L, the event horizon,

$$L \cong Gm/c^2.$$

On the other hand, the uncertainty principle places this limit at the Compton wavelength λ_c:

$$\lambda_c = h/mc.$$

Equating these yields the Planck mass $m_p = (hc/G)^{1/2} = 5.5 \times 10^{-8}$ kg. The length $L = \lambda_c = (Gh/c^3)^{1/2} \approx 10^{-35}$ m is called the Planck length, and the time for light to travel across that length,

$$t_p = \lambda_c/c = Gh/c^5 = 1.35 \times 10^{-43} \, s,$$

is called the Planck time. Because general relativity breaks down for times earlier than the Planck time, no one knows how to describe the universe before Planck time. Relativistic space-time is no longer a continuum, and a new theory of gravity, quantum gravity, or supergravity, is needed.

11.2 Cosmic Background Radiation

We have no way of knowing directly the physical conditions of the Big Bang epoch. We can only look for relics of the Big Bang. George Gamow did pioneering work in this field in the mid-1940s. Gamow was interested in the origin of elements. Starting from the basic building blocks of protons and neutrons, he attempted to describe the formation of nuclei through the fusion process. Astrophysicists already knew by the 1940s that such fusion processes operate inside stars, where the necessary conditions of high temperature and density were known to exist. Gamow pointed out that similar conditions must have existed in a typical Friedmann universe soon after the Big Bang.

We know from (8.40) that the mass density $\rho(t)$ of the matter in the universe was very high at small values of R:

$$\rho(t) = \frac{\rho_o R_o^3}{R^3(t)} \tag{11.1}$$

where the subscripts 0 indicate present values. A simple calculation shows that the temperature was also very high at small R. The early universe contains radiation in the form of photons moving in all directions with very high frequencies. Expansion causes the radiation energy density $\varepsilon_r(t)$ to decrease as $R^{-4}(t)$ for two reasons: first, the number density of photons decreases as $R^{-3}(t)$ (because volumes expand as $R^3(t)$); second, the energy of individual photons decreases as $R^{-1}(t)$ (because of the redshift in frequency). Therefore $\varepsilon_r(t)$ decreases as $R^{-4}(t)$, and

$$\varepsilon_r(t) = \frac{\varepsilon_r(t_o) R_o^4}{R^4(t)} \tag{11.2}$$

where $\varepsilon_r(t_o)$ is the present value of a radiation energy density that is a relic of an early hot era. The equivalent mass density $\rho_r(t)$ is

$$\rho_r(t) = \varepsilon_r(t)/c^2 \tag{11.3}$$

and therefore also decreases as $R^{-4}(t)$. Therefore, as we go backward in time the radiation density increases faster than the matter density, so that, however small ε_r is now, there must have been a time t_E when the densities of radiation and matter were equal. At such a time we have

$$\rho(t_E) = \rho_r(t_E), \qquad (\rho_o R_o^3)/R^3(t_E) = (\rho_r(t_o) R_o^4)/R^4(t_E)$$

which gives

$$\rho_r(t_o)/\rho_o = R(t_E)/R_o. \tag{11.4}$$

At times earlier than t_E, ρ_r was greater than ρ, and radiation was more dominant. Gamow therefore assumed that in the early epochs the dynamics of expansion were determined by radiant energy. The period $0 < t < t_E$ is the radiation-dominated era of the history of the universe. The contents of the radiation-dominated universe are often referred to as the *primeval fireball*, or as *ylem* (from the Greek *hyle*, meaning "that on which form has yet to be imposed").

There is good reason to believe, as we shall see later, that radiation and matter (in the form of relativistic particles and antiparticles) were in thermal equilibrium at the same temperature T, and that the radiation had a blackbody spectrum. As the universe expanded, the mean photon energy decreased as $R^{-1}(t)$ because of the redshift, so that the temperature T, which is proportional to this mean energy, should also have decreased as $R^{-1}(t)$. Since the radiation preserves its blackbody character, T continues to be proportional to $R^{-1}(t)$. Thus, if T_o is the present radiation temperature, then

$$T(t) = \frac{R_o}{R(t)} T_o, \tag{11.5}$$

which says that the temperature was very high at small values of R. We can carry the calculation one more step to get an explicit formula that relates radiation temperature T to time t. Since the radiation was in blackbody form with temperature T,

$$\varepsilon_r = aT^4 \tag{11.6}$$

where a is the radiation constant ($= 7.5 \times 10^{-16}\,\mathrm{J\,m^{-3}\,K^{-4}}$). This means that in the early universe

$$T_o^o = aT^4, T_1^1 = T_2^2 = T_3^3 = -\frac{1}{3}aT^4. \tag{11.7}$$

The curvature parameter k, from (8.43) and (8.45) when applied to the present epoch, is given by

$$k = (2q - 1)H_o^2 R_o^2/c^2 = (2q - 1)\dot{R}_o^2 c^{-2}. \tag{11.8}$$

Since k will not affect the dynamics of the early universe significantly, we set it equal to zero. Thus, (8.32) becomes

$$\frac{\dot{R}^2}{R^2} = \frac{8\pi Ga}{3c^2} T^4.$$

Substituting (11.5) into the above equation, we get

$$\dot{R} = A^2 \left(\frac{8\pi Ga}{3c^2}\right)^{1/2} \frac{1}{R}; \quad A = R_o T_o,$$

which can be easily integrated to get

$$R = A \left(32\pi Ga/3c^2\right)^{1/4} t^{1/2} \tag{11.9}$$

where we have employed the initial condition: $R = 0$ at $t = 0$. Eliminating R from (11.9) and (11.5) we obtain the desired result

$$T = \left(3c^2/8\pi Ga\right)^{1/4} t^{-1/2}. \tag{11.10}$$

All of the quantities inside the parentheses on the right hand side are known physical quantities. Thus we obtain, after substituting numerical values for all the known physical quantities

$$T(^o K) = 1.52 \times 10^{10} t^{-1/2} (\text{sec}), \tag{11.11}$$

and we can see that about one second after the Big Bang the radiation temperature of the universe was on the order of $10^{10} K$. And at Planck time ($t \sim 10^{-43}$ s), $T \sim 10^{32} K$, which is close to the maximum temperature T of blackbody radiation found by A. Sakharov in 1966: $T_{\max} = (\alpha/k)(\hbar c^5/G)^{1/2}$, where α, $\hbar$, c, and G denote, respectively, a constant factor near unity, the Boltzmann constant, the Planck constant, the speed of light, and the gravitational constant. Substitution of these constants gives $T_{\max} \sim 10^{32} K$. Sakharov's deduction starts from the thermodynamic properties of the hot matter in the isotropic universe within the framework of gravitational perturbation theory (see the reference at the end of the chapter).

The idea of a hot Big Bang depends on the basic assumption that the primeval radiation had a blackbody spectrum and that it preserved its blackbody character as the universe expanded. And there should be relic radiation present today. We have addressed, in Chapter 7, the question of why we believe the primeval radiation had a blackbody spectrum. Let us now discuss this important question again in a different way.

Immediately following the Big Bang, the universe was so incredibly hot that all matter behaved like photons, since particles (whatever they may be!) move essentially at the velocity of light. Hence the initial state was a chaotic, gaseous inferno of high-energy photons and elementary particles. We take up the story from the state when baryons (protons and neutrons), leptons (electrons, muons, neutrinos, and their antiparticles), and photons are already in existence. These particles would interact and collide, but only for very brief time spans, and so their effects on motions may be otherwise neglected. That is, these particles would act as particles of an ideal gas. However, the collisions and scatterings of the particles would have helped to redistribute their energies and momenta. The radiation (photons) therefore stayed in thermal equilibrium with the matter at the same temperature and had a blackbody spectrum.

One of the characteristics of cosmic blackbody radiation is that the total number of photons is conserved. The dominant reactions, at the end of the hadron era, are electromagnetic, $\gamma + \gamma \rightarrow e^+ + e^-$ and bremsstrahlung such as $e \rightarrow e + \gamma$ and $e + \gamma \rightarrow e$ in the presence of an external field. These processes are in equilibrium so that charge conservation requires that the number of photons is also conserved.

Our next task is to show that the radiation should preserve its blackbody character. The proof is based on the conservation of photon number. Now the number $dN(t)$ of photons in the frequency band v and $v + dv$ in a volume $V(t)$ of space at cosmic time t is given by Planck's law:

$$dN(t) = \frac{8\pi v^2 V(t)dv}{c^3 \left[\exp(hv/kT(t)) - 1\right]}.$$ (11.12)

As time proceeds, the number of photons in the volume remains the same, because of the conservation of photon number. This means that a co-moving observer would see as many photons crossing the (imaginary) boundary into this volume as leaving it. At a new time t', the original group of photons has been redshifted to a frequency

$$v' = \frac{vR(t)}{R(t')}, \quad dv' = dv\frac{R(t)}{R(t')}$$ (11.13)

while the volume has expanded to

$$V(t') = V(t)\frac{R^3(t')}{R^3(t)}.$$ (11.14)

Therefore, for this group of photons, we have

$$dn(t') = dN(t) = \frac{8\pi V(t')v'^2dv'}{c^3(\exp(hv'/kT(t')) - 1)}.$$ (11.15)

This looks just like a blackbody spectrum at a new temperature $T(t')$, where $T(t')$ is

$$T(t') = [R(t)/R(t')]T(t).$$ (11.16)

Thus, the radiation keeps its blackbody character but will appear cooler by factor $R(t)/R(t')$. We should expect to see a low temperature relic radiation background. Gamow and his co-workers predicted in 1948 that this relic radiation should currently have a temperature of about 5 K. A blackbody radiation with a temperature of this order is predominantly in the microwave form. Gamow's prediction was not taken seriously. But as we learned in Chapter 7, this relic radiation was indeed detected accidentally by Penzias and Wilson in 1965. The present value of the energy density of this cosmic microwave background is

$$\varepsilon_r(t_o) = aT^4 = (7.5x10^{-16}Jm^{-3}K^{-4}) \times (2.7K)^4 = 4 \times 10^{-14}Jm^{-3}.$$

The equivalent mass density, $\rho_r = \varepsilon_r/c^2$, is

$$\rho_r(t_o) = 4.5x10^{-31}kgm^{-3}.$$ (11.17)

There are many photons of starlight traveling across space, and this is not distributed with a blackbody spectrum. The total non-blackbody energy density has been estimated as less than $\varepsilon_r(t_0)/100$.

Note that the radiation mass density given by (11.17) is only about a thousandth of the value for the matter density ρ_0. Hence the universe is now matter-dominated and in the matter era. But at some time in the past, the energy density of radiation exceeded that of matter, so the universe was radiation-dominated and in the radiation era. To see this, let us shrink the universe. Then the matter density increases as

$$\rho_m \propto R^{-3}.$$

In contrast, the radiation density goes as T^4. Now, the wavelength of a photon is proportional $R(t)$

$$\lambda \propto R$$

and because a photon's energy is $E = h\nu = hc/\lambda$,

$$E \propto R^{-1}$$

and for blackbody radiation,

$$T \propto R^{-1}$$

so that

$$\rho_r \propto R^{-4}.$$

Thus, at some time in the past, the energy density of radiation exceeded that of matter and the universe was radiation-dominated.

11.2.1 Conservation of Photon Numbers

Since the volume of a co-volume increases with the cube of $R(R^3)$, the particle density (number of particles per unit of volume) decreases inversely with the cube of R (R^{-3}). In addition, the number of photons in fossil radiation is proportional to the cube of the temperature (T^3), and this temperature decreases inversely proportion to the expansion (T is proportional to R^{-1}). Therefore, the number of fossil radiation photons in a co-volume (the product of the number of photons per unit of volume multiplied by the volume of the co-volume) remains constant over the course of time: $R^3 \times R^{-3} = 1$.

Alternatively, we can use (11.16). As the universe expands, the cosmic background radiation keeps its blackbody character, but it appears cooler by a factor $R(t)/R(t')$, and the new temperature is given by (11.16), now written as:

$$T_0 = [R/R_0]T;$$

or

$$T_0 R_0 = RT \qquad (11.18)$$

where we have replaced $T(t')$ and $R(t')$ by the present values T_0 and R_0, respectively. R and T stand for $R(t)$ and $T(t)$. Now, the number density of photons in a blackbody distribution is proportional to T^3, and the total number of photons in the universe is proportional to $T^3 R^3$, which is equal to $T_0^3 R_0^3$ and is constant. Thus the total number of photons in the universe is conserved as the universe expands.

11.2.2 The Transition Temperature T_t

When the universe was dominated by radiation, the radiation and matter were in thermal equilibrium at the same temperature. At some time t_E, the universe entered the matter-dominated era and the radiation temperature was no longer equal to the matter temperature. We can make an estimate of the transition temperature T_t. From (11.4), if we take $\varepsilon_r(t_0) = 4.5 \times 10^{-31}$ kg m^{-3} and $\rho_0 = 3 \times 10^{-28}$ kg m^{-3}, we get

$$R(t_o)/R(t_E) = 700$$

where we have used the lowest estimate of ρ_0. If dark matter exists, then $R(t_0)/R(t_E)$ will be greater than 700. From (11.5) and if we take the value 2.7 K for the present radiation temperature, we get the transition temperature T_t

$$T_t = 1900 \,\text{K}. \tag{11.19}$$

Again, if there is dark matter T_t will be greater than this. Now this value of 1900 K is of the same order as 4000 K, above which a dilute gas of hydrogen would be almost completely ionized. Thus, as we shall see later, the transition period between the radiation and matter-dominated eras, $t > t_E$ is also the recombination era during which the plasma of free electrons and protons condenses into neutral hydrogen, which is much less opaque to radiation than plasma.

11.2.3 The Photon-to-Baryon Ratio

The observed universe contains about 10^9 photons for every proton or neutron. The photons are mainly in the cosmic background radiation, whereas the protons and neutrons form the atomic nuclei of the matter that makes up the galaxies. The standard Big Bang theory does not explain this ratio but instead assumes that the ratio is given as a property of the initial conditions.

The Russian physicist Andrei Sakharov first suggested the idea that particle physics could provide an answer to this question. More detailed calculations, in the context of grand unified theories, were carried out by Yoshimura of Tohoku University in Japan and by Weinberg in the United States. This was the first application of grand unified theories to cosmology, and the subject remains crucial to our understanding of cosmology in this context.

All physical processes observed up to now obey the principle of baryon number conservation: the total baryon number of an isolated system cannot be changed. In the early universe protons and neutrons rapidly interconverted, by processes such as

$$\text{proton} + \text{electron} \to \text{neutron} + \text{neutrino}.$$

The baryon number is left unchanged by the reaction above, since the proton and the neutron each have a baryon number of 1. Similarly, at high energies the reaction

$$\text{electron} + \text{positron} \to \text{proton} + \text{antiproton}$$

is frequently observed. The total baryon number on the left side is 0; it is also 0 on the right: $1 + (-1) = 0$. So the principle of baryon number conservation holds.

To estimate the baryon number of the observed universe, we must know whether all the distant galaxies are composed of matter, or whether some of the distant galaxies might be formed from antimatter. We do not know the definite answer to this yet, but there is a strong consensus that the observed universe is probably made of matter. This consensus is motivated by the absence of any known mechanism that could have separated matter from antimatter over the large distances that separate galaxies. Assuming that this belief is true, then the ratio of photon to baryon is about 10^{10}; i.e., the observed universe contains about 10^{10} photons for every baryon. The estimate is simple: The energy density associated with blackbody radiation of temperature T is aT^4, and the mean energy per photon is $\sim kT$, so the number density of blackbody photons is, for $T = 2.7\,\text{K}$:

$$N_{ph} = aT^4/kT = aT^3/k = 3.7 \times 10^2 \text{ photons/cm}^3 \qquad (11.20)$$

where $a = 7.56 \times 10^{-15} \text{ erg cm}^{-3} \text{ k}^{-4}$, $k = 1.38 \times 10^{-16} \text{ erg K}^{-1}$. The number density of baryons equals ρ_m/m_p, where m_p is the mass of the proton ($= 1.66 \times 10^{-24}$ g) and ρ_m is the mass density of the universe. If we take $\rho_m = \rho_o$ (the critical density) $= 3 \times 10^{-31} \text{ g cm}^{-3}$, we find the number density of baryons is $\sim 0.22 \times 10^{-6}$ baryons/cm^3. Thus, the baryon/photon ratio is approximately equal to 10^{-10}:

$$N_{ph}/N_{ba} = 3.7 \times 10^2/0.22 \times 10^{-6} \cong 10^{10}. \qquad (11.21)$$

The total number of photons is conserved as the universe expands. Similarly for matter, the baryon number is also conserved. So the above ratio of photons to baryons will stay constant as the universe expands.

11.3 The Creation of Matter and Photons

So, where did all the matter and radiation in the universe come from? Recent intriguing theoretical research by physicists such as Zeldovich, Weinberg, and Guth suggest that the universe might have started as a perfect vacuum and that all the particles of the material world were created from the expansion of space-time. The detailed calculations are complex, but the basic ideas are not difficult to understand with the help of Heisenberg's uncertainty principle and the concept of quantum fluctuation. We have seen how the uncertainty principle and the concept of quantum fluctuation can help us to understand the Hawking effect. As we know, the vacuum of classical physics is an uninteresting empty state. However, quantum vacuum is a very interesting place, where energy exists but particles do not. Fluctuations of quantum vacuum can lead to the temporary formation of particle-antiparticle pairs. Normally, the mass Δm of the pair and its lifespan Δt must satisfy the "Heisenberg-Einstein" relation:

$$\Delta m x \Delta t \geq h/(2\pi c^2) \tag{11.22}$$

so that annihilation follows closely upon creation and the pairs are virtual (Fig. 11.1a). But during the Big Bang, space-time was expanding so fast that the particle and antiparticle might have been pulled apart and gained real existence (Fig. 11.1b).

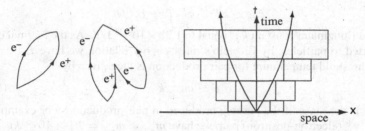

Fig. 11.1 (a) Creation and annihilation (b) In expanding space-time.

Were photons also created from the space-time expansion? The answer is no. Quantum mechanics forbids the direct creation of photons from the vacuum; thus, particles have to be created first. We now make some order of magnitude estimate for the particle creation from the expansion of space-time.

Suppose at the end of creation the particles are pulled apart by a distance d in a time interval Δt, and if the relative acceleration during the separation is g, then

$$d \sim g(\Delta t)^2. \tag{11.23}$$

If gravity is the only force field in the early universe, we may write

$$g \sim (\ddot{R}/R)d \sim d/t^2. \tag{11.24}$$

Note that here t is the age of the universe. Substituting (11.24) into (11.23) shows $\Delta t \sim t$, i.e., the interval over which the particles can be pulled apart is comparable to the age of the universe. The particles move essentially at the speed of light, so we may write d $\sim ct$. Now, with $\Delta t \sim t$ and $d \sim ct$, (11.22) implies that, at the end of the process

$$(2m)c^2 t = (2m)c(ct) \approx (2m)cd \approx h/2\pi$$

from which we have

$$d \approx h/mc \tag{11.25}$$

where, for order of magnitude estimate, we have dropped the factor 2 and π. Thus, the creation of particles of mass m occurs primarily when the cosmic time equals the Compton time $h/mc^2 : t \sim d/c = h/mc^2$. At the end of the process, the pair is pulled apart by a distance d approximately equal to the Compton wavelength h/mc. Roughly, one particle would be created per Compton volume $d^3 \sim (h/mc)^3$.

The creation of pairs of particle-antiparticle in the early epochs of the universe depends on the temperature of the radiation. The critical temperature at which particles of a given type can be spontaneously produced is called the threshold temperature for that type of particle. Near this threshold temperature, collisions between particle and antiparticle pairs produce high-energy γ-ray photons that can be converted into particle and antiparticle pairs again. A thermal equilibrium will be reached very soon. When the temperature drops below the threshold temperature for a particular particle, particle and antiparticle pairs cannot be created. To estimate the threshold temperature for pair production, we first note that the thermal energy is approximately given by

$$E_{th} \sim kT \quad \text{or} \quad T \sim E_{th}/k$$

where the Boltzmann constant k is equal to 1.38×10^{-23} J/K. As the thermal energy is converted to particles, by Einstein's mass-energy relation we have $E_{th} = mc^2$, and the threshold temperature for pair production is then given by

$$T \sim mc^2/k \tag{11.26}$$

where m is the mass of the particle produced in pair production. For example, for the $e^- - e^+$ (electron-positron) pair, we have $m_{e^-} = m_{e^+} = 9.1 \times 10^{-31} kg$, and

$$E = (m_{e^-} + m_{e^+})c^2 = 2 \times 9.1 \times 10^{-31} kg \times (3 \times 10^8 m)^2 = 1.638 \times 10^{-13} J,$$

so the threshold temperature for $e^- e^+$ production is $T \sim E/k = 1 \times 10^{10}$ K.

For a proton-antiproton $(p - \overline{p})$ pair, the threshold temperature is

$$T \sim 1836 \times 10^{10} \text{ K} = 2 \times 10^{13} \text{ K},$$

as $m_p = 1836 m_e$. It is also the threshold temperature for neutron-antineutron pair, as $m_n = 1838 m_e$, almost equal to the mass of proton.

In a similar way we can calculate the threshold temperatures for other types of particle-antiparticle pairs. The higher the mass of the particle, the greater temperature is required. Threshold temperatures of several common types of particles, with their masses, are listed in Table 11.1, where $1 \text{MeV} = 10^6$ eV. 1 eV is the energy

Table 11.1 Threshold Temperature of Selected Particles

Particles	Rest Energy (MeV)	Threshold Temperature (10^9 K)
Neutrino ν	0.00001(?)	0.00001
Electron $e^{\pm}$	0.5110	5.930
Muon $\mu^{\pm}$	105.55	1, 226.2
Pion $\pi^{\pm}, \pi^0$	134.96/139.57	1, 556.2
Proton p	938.26	10, 888
Neutron n	939.55	10, 930

gained by a charge e that has been accelerated through a potential of 1 volt, which is equal to $(1.6 \times 10^{-19}C)(1V) = 1.6 \times 10^{-19}$ J. 1 eV also corresponds to 5040.2 K, or $1 K = 0.8617 \times 10^{-4}$ eV.

11.4 A Brief History of the Early Universe

To attempt to give a rough outline of the evolution of the very early universe is an ambitious task. Some of the ideas involved are still very tentative. Interested readers should read the classical book on this subject by Steven Weinberg (see the references at end of this chapter).

11.4.1 The Planck Epoch

We can say very little about the first 10^{-43} s immediately after the Big Bang, as here we enter a truly alien domain. At the moment of the Big Bang, space-time is completely jumbled up in a state of infinite curvature like that at the center of a black hole. Thus, we should think of the Big Bang as an explosion of space at the beginning of time. We cannot use the existing laws of physics to tell us exactly what happened at the moment of Big Bang and what existed before the Big Bang. Without a clear background of space-time, concepts such as past, future, and here and now cease to have meaning. This short time interval, called the Planck time (t_P), lasted only for about 10^{-43} seconds:

$$t_P = \sqrt{Gh/c^5} = 1.35 \times 10^{-43} s,$$

where G is the gravitational constant, h the Planck constant, and c the speed of light. From the Big Bang to the Planck time 10^{-43} s later, physics fails us. Even general relativity breaks down for times earlier than the Planck time; no one knows how space-time and matter behaved during the Planck epoch. Nevertheless, J. A. Wheeler has speculated that space-time as we know today burst forth from a seething, foam-like, space-time mishmash during the Planck time.

During the Planck era, the four basic forces—strong, electromagnetic, weak, and gravity—were united as a single force. According to quantum field theory, a force is transmitted by the exchange of innumerable force-carrying particles, the gauge bosons. Each of the four basic forces today in nature is carried by a specific gauge particle. The electromagnetic force is transmitted by gauge bosons called photons. The strong force is carried by gauge bosons called gluons, and the intermediate vector bosons, $W^{\pm}$ and Z^0 particles, carry the weak force that causes the decay of unstable nuclei. Similarly, physicists believe that gauge bosons called gravitons carry gravity; this quantum property of gravity should become very effective at Planck length scale. During the Planck era, all of the four natural forces would be indistinguishable from one another; only one kind of gauge boson, the graviton, dominated the activity.

In the language of general relativity, gravity is a consequence of the deformation of space caused by the presence of matter and energy. So in quantum gravity theory, gravitons represent individual packages of curved space-time that travel at the speed of light. The appearance and disappearance of innumerable gravitons give the geometry of space a lumpy, ever-changing appearance. Wheeler thinks of it as foam substructure where the geometry of space twists and contorts. Thus, even our concepts of space and time have to be reevaluated in the face of the quantum fluctuations of space-time in the Planck era. But the effects of quantum gravity are completely undetectable at the atomic and nuclear scale. For a comparison, the difference in size between an atom and the domain of the graviton is proportionally the same as that between the sun and an atom.

The Russian physicists Ya. Zel'dovitch and A. Starobinski proposed, in the early 1970s, that the changing geometry of space during the Planck era might actually have created all the matter, antimatter, and radiation that exist. In their picture of creation, the rapidly changing geometry of space created massive particles and antiparticles. The production of matter and antimatter removed energy from the enormous fluctuations occurring in the geometry of space and, by the end of the Planck era, succeeded in damping them out altogether. Their calculations also showed that the rate of particle creation increased as more and more particles were created.

Several recent studies by physicists Edward Tryon, R. Brout, F. Englert, E. Cunzig, David Atkatz, and Heinz Pagels have shed additional light on the Big Bang. Imagine, if you can, nothing at all–this is the primordial vacuum of space. In this infinite emptiness, random fluctuations in the very geometry of space ever so slightly changed the energy of the vacuum at various points. Eventually, one of these fluctuations attained a critical energy and began to grow. As it grew, the massive leptoquarks and antileptoquark particles were created; expansion accelerated, creating more leptoquarks. This furious cycle continued until, at long last, the leptoquarks decayed into quarks, leptons (particles like electrons and muons, for example), and their antiparticles. The fluctuations in the geometry of space subsided once the universe emerged from the Planck era, 10^{-43} second after its birth. The density of quarks and gluons everywhere was higher than that inside a proton today. The universe was filled with plasma of all possible types of fundamental particles.

So we are left with the remarkable possibility that, in the beginning, there existed nothing at all, and that nearly all of the matter and radiation we now see emerged from it. Physicist Frank Wilczyk has described this process: "The reason that there is something instead of nothing," he says, "is that 'nothing' is unstable." A ball sitting on the summit of a steep hill needs but the slightest tap to set it in motion. A random fluctuation in space is apparently all that was required to unleash the incredible latent energy of the vacuum, creating matter and energy and an expanding universe from, quite literally, nothing at all.

The universe did not spring into being instantaneously, but was created a little bit at a time. Once a few particles were created by quantum fluctuations of the empty vacuum, it became easier for a few more to appear, and so forth. In a rapidly escalating process, the universe gushed forth from nothingness. The primordial vacuum could have existed for an eternity before the particular fluctuation that gave rise to our universe occurred. Or, as physicist Edward Tryon puts it, "Our universe is simply one of those things that happens from time to time."

The principles of quantum gravity may ultimately force us to reconsider questions such as, "What happened before the Big Bang?" Perhaps the complete theory of quantum gravity will tell us how to ask the right questions.

After the Planck era, space and time began to behave in the way we think of them today. We can safely apply the laws of physics to study the early universe.

11.4.2 The GUTs Era

As soon as the gravitational force is frozen out by the cooling universe at about 10^{-43} s, the Planck era ended. At this point the temperature was slightly less than 10^{32} K and the average energy of the particles was about 10^{19} GeV. The strong, weak, and electromagnetic forces were all still indistinguishable from one another and united together as a single force. The theories that describe this unified force are known collectively as Grand Unified Theories, or GUTs for short. Accordingly, we refer to this period of time as the GUT era. According to Sakharov, during this period, quantum numbers were not conserved, and a slight excess of quarks over antiquarks occurred, roughly 1 in 10^9, that ultimately resulted in the matter that we now observe in the universe.

Temperature played a crucial controlling role in the evolution of the early universe. As shown earlier, the temperature of the universe at any given time after Planck time is given approximately by (11-11), which tells us that the temperature of the universe during the GUTs era was still incredibly high, around 10^{28} K. The size of the universe also had an abrupt change, due to the "freezing out" of gravity–a "phase transition," as we discussed in Chapter 10. The GUT era represents a period when the universe underwent a "phase transition" from a higher energy state to one of lower energy. This is analogous to a ball rolling down the side of a mountain and coming to rest in the lowest valley. As the universe "rolled downhill," it began a brief but stupendous period of expansion. The universe swelled to billions of times its former size in almost no time at all.

11.4.3 The Inflationary Era

At 10^{-35} s, the universe had expanded sufficiently to cool to about 10^{27} K, at which point another phase transition occurred as the strong force condensed out of the GUTs group, leaving only the electromagnetic and weak forces still unified as the electroweak force. The released latent energy during the phase transition became the dynamite for the universe to undergo an extraordinarily rapid inflationary expansion, as explained in Chapter 10. The universe tripled its size every 10^{-34} s. Although the inflationary epoch lasted for only a fraction of a second (from 10^{-35} s to

10^{-33} s), the universe increased its size by approximately 10^{50} times and consisted of a quark soup coexisting with leptons interacting via the electroweak interaction. Baryon nonconserving processes at this era would have resulted in the net excess of baryons over antibaryons.

11.4.4 The Hadron Era

The hadron era roughly covers the period 10^{-35} s $< t < 10^{-6}$ s. During this era the universe consisted of a soup of quarks and leptons and their antiparticles in thermodynamic equilibrium. As the universe continued to expand and cool adiabatically, at a temperature of about 10^{16} K($t \sim 10^{-12}$ s), the $W^{\pm}$ and Z^0 bosons behaved like massive particles and the photons had no mass. The weak and the electromagnetic forces began to display their separate characteristics and broke their symmetry. The universe underwent another phase transition and was then populated with electron-quark plasma.

When the temperature of the universe cooled to about 10^{13} K($t \sim 10^{-6}$ s), quarks and their antiparticles annihilated each other, and the residues combined to form protons and neutrons in equal numbers. For $kT > M_{hadron}c^2$, there are many hadrons and antihadrons in the hot plasma. As these particles have short lifetimes, they were frozen out very quickly. In Table 11.2, we list some of the common hadrons (the nucleons and the pions), their masses, and their lifetimes. As kT drops below

$$KT \sim m_\pi c^2 \sim 140\,\mathrm{MeV} \quad \text{or} \quad T \sim 10^{13}\,K.$$

The pion, the lightest hadron, can no longer be produced. Through annihilation and decay, it will soon disappear, so that the only hadrons remaining are the stable protons and the relatively stable neutrons, in thermal equilibrium with the photons and the weakly interacting leptons. One of the interesting problems not fully understood yet is why matter, in the form of baryons p and n, has been favored over antimatter, the antibaryons, $\overline{p}$ and $\overline{n}$. According to Sakharov, this is likely due to a slight asymmetry in the fundamental laws (the nonconservation of baryon numbers) that caused the number of quarks formed originally to exceed the number of antiquarks by 1 part in 10^9.

By about 0.1 milliseconds (10^{-4} s), the temperature had dropped to about 10^{12} K, muons and antimuons annihilated each other, and muon neutrinos and antineutrinos decoupled from everything else.

Table 11.2 Some of the common hadrons

	Particles		Spin	Mass (MeV/c^2)	Lifetime (sec)
Hadron	Baryons	p	1/2	938.28	stable
		n	1/2	939.57	10^3
	Mesons	$\pi^{\pm}$	0	140.	10^{-8}
		π^0	0	135.	10^{-16}

When $T < 10^{11}$ K($t \sim 10^{-2}$ s), the neutron-proton mass difference (= 1.3 MeV/c^2) began to shift the neutron-proton ratio through equilibrium of the weak interaction processes

$$n + \nu_e \leftrightarrow p + e$$
$$n + e^+ \leftrightarrow p + \bar{\nu}_e$$
$$n \leftrightarrow p + e^- + \bar{\nu}_e.$$

The ratio of the number of neutrons to the number of protons depended on the temperature according to

$$\frac{N_e}{N_p} = \exp\left(-\frac{1.52 \times 10^{10}}{T}\right).$$

As the temperature falls below the threshold temperature for the creation of protons and neutrons, the neutrons and protons are frozen out and the hadron era ends. Now the dominant reactions are electromagnetic, $\gamma + \gamma \rightarrow e^- + e^+$ and bremsstrahlung such as $e \rightarrow e + \gamma$ and $e + \gamma \rightarrow e$. These processes are in equilibrium so that charge conservation requires that the number of photons is also conserved. An important result, the ratio of photons to baryons is established at about 10^9, as shown earlier.

11.4.5 The Lepton Era

The equilibrium process resulting from weak interaction $p + e^- \rightarrow n + \nu_e$ and $n + e^+ \rightarrow p + \bar{\nu}_e$ now becomes the principal process at work. At $t \sim 2$ s, the temperature fell sufficiently to prevent the interaction $n + \nu_e \rightarrow p + e^-$ and $p + \bar{\nu}_e \rightarrow n + e^+$ from taking place any longer; electron neutrinos and antineutrinos start to decouple from everything, and form a cosmic neutrinos background permeating the universe today.

The equilibrium of leptons and photons, resulting from the interactions $e^- + e^+ \rightarrow \gamma + \gamma$ and $\gamma + \gamma \rightarrow e^- + e^+$, is maintained until $t \sim 4$ s, when the temperature fell below 6×10^9 s and the process $\gamma + \gamma \rightarrow e^- + e^+$ is no longer energetically possible. Consequently, the electrons and positrons annihilated, leaving a small excess of electrons. This, together with the cooling of the neutrinos during the expansion of the universe, puts an end to the weak interaction processes, with the exception of β-decay ($n \rightarrow p + e^- + \bar{\nu}_e$).

The lepton era ends at about $t = 10$ s ($T \sim 10^9$ K and $kT \sim 1/2$ MeV). Further expansion and cooling dropped the average photon energy below that needed to form electron-positron pairs.

11.4.6 The Nuclear Era

At about $t \approx 3$ min and $T \approx 10^9$ K, nucleosynthesis begins to dominate over nuclear break-up collisions. When the temperature is higher than 10^9 K, any deuterium nucleus formed by the process $p + n \rightarrow {}^2H$ is immediately destroyed by photodisintegration $\gamma + {}^2H \rightarrow P + n$, since the binding energy of the deuterium is only 2.2 MeV. Although temperatures and densities are certainly high enough for fusion to occur, the process cannot get under way because deuterium is destroyed as fast as it appears. The universe has to wait until it becomes cool enough for the deuterium to survive. This waiting period is sometimes called the deuterium bottleneck.

When the temperature falls below 10^9 K, photodisintegration is no longer possible and deuterium is at last able to form and endure. The abundance of deuterons then climbs swiftly.

Deuterons react swiftly with protons, and a series of nuclearlike reactions convert deuterium into heavier elements:

$$n + p \rightarrow {}^2H + \gamma, \quad {}^2H + {}^2H \rightarrow {}^3He + n$$
$$ {}^3He + n \rightarrow {}^3H + p, \quad {}^3H + {}^2H \rightarrow {}^4He + n.$$

For about 200 seconds, the temperature remains high enough for nuclear reactions to change the chemical makeup of the universe from entirely hydrogen (protons) into a more complex mixture that includes protons (^1H), deuterons (^2H), 2 isotopes of helium (^3He and ^{4}He), and small amounts of lithium (^7Li) and beryllium (^7Be). The way that these nuclei change with time is shown in Fig. 11.2 (the detailed calculation was done by Robert V. Wagoner of Caltech), which shows a universe whose matter content is primarily hydrogen and helium.

Eventually, the density of neutrons gets too low, and the time between collisions with protons become longer, with the fusion processes freezing out. Unstable nuclei can still decay, but the stable nuclei such as helium-4 and lithium-7 produced in this period are around today.

By the time the universe is about 15 minutes old, much of the helium we observe today has been formed, and the universe becomes too cool for further fusion to continue. The formation of heavier elements has to await the birth of stars. In stars, the density and the temperature both increase slowly with time, allowing more and more massive nuclei to form, but in the early universe the opposite is true. The temperature and density are both decreasing rapidly, making conditions less and less favorable for fusion as time goes on.

Careful calculations indicate that by the end of the nuclear era about 1 helium nucleus had formed for every 12 protons remaining. Since a helium nucleus is four times more massive than a proton, helium accounted for about one quarter of the total mass of matter in the universe:

$$\frac{1 \text{ helium nucleus}}{12 \text{ protons} + 1 \text{ helium nucleus}} = \frac{4 \text{ mass units}}{12 \text{ mass units} + 4 \text{ mass units}} = \frac{4}{16} = \frac{1}{4}.$$

The remaining 75% of the matter in the universe was hydrogen. If all neutrons were gone before the nuclei were stable in a typical collision, then only hydrogen would be produced and the universe would be a very different world today.

Until about 700,000 years after the Big Bang, the universe was radiation dominated. Now the universe grew to about 1/1,000 of its present size, with its temperature down to about 3000 K. This temperature was low enough for protons to combine
with electrons, forming neutral hydrogen atoms via the reaction $e^- + p \rightarrow H + \gamma$. At
this point the scattering of photons from neutral hydrogen (as opposed to free protons and electrons) dropped dramatically, and electromagnetic radiation became free
to pass throughout the universe. From this time on, matter dominated the universe,
with more energy being in the form of matter than radiation. Photons now became
free to pass throughout the universe; a blackbody radiation of temperature 3000 K
should persist forever. However, this background radiation characteristic of 3000 K
had redshifted steadily as the expansion and cooling continued, and we measure the
vast majority of photons in the background radiation today to be about 2.7 K. This
is the cosmic microwave background radiation. Atoms are now able to form, and
matter begins to clump together to form molecules, gas clouds, stars, and eventually
galaxies.

The primordial abundances of these light elements have been confirmed by observations. The theory of stellar nucleosynthesis accounts very well for the observed
abundances of heavy elements in the universe. But there are some conflicts between theory and observations when it comes to the abundances of the light elements helium, lithium, beryllium, and boron. Simply put, there appears to be more
of those elements than can be explained by nuclear fusion in stars over the lifetime
of the galaxy. What is more, astronomers find that, no matter where they look and no
matter how low a star's abundance of heavy elements, there seems to be a minimum
amount of helium—between 20% and 25% by mass—in all stars. The most obvious

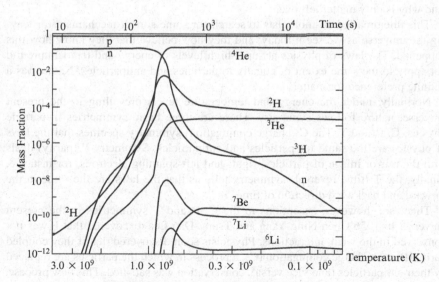

Fig. 11.2 Nucleosynthesis in the early universe.

explanation is that this base level of helium is primordial; that is, formed during the early, hot epochs of the universe, as we described above.

The relative abundances of light elements in primordial material give us an important clue about conditions in the early universe. These are the kinds of abundances that would have resulted from a low-density environment rather than a high-density environment. In fact, the most probable value for the density of the universe at the time nuclear reactions were taking place was only a few percent of the critical density. Thus, the abundances of the isotopes produced during the first few minutes of the Big Bang indicate that matter in the form of electrons and nuclei falls far short of providing the self-gravity to produce a positively curved, finite universe. If there is enough "dark" matter in the universe to produce positive curvature, it can't be electrons and nuclei, but must take the form of exotic, perhaps undiscovered, types of particles.

11.5 The Mystery of Antimatter

The creation of particle-antiparticle pairs from energy opens the way to explaining where the material of the universe came from. However, the pair-creation should produce equal amounts of particles and antiparticles. As the universe rapidly expanded and cooled, this explosive mixture should have undergone wholesale annihilation as positrons ran into electrons, protons into antiprotons, and neutrons into antineutrons. The result should be a universe populated not by atoms but gamma rays. Yet the universe has considerable matter in it. Why is it not empty of matter? And why is only matter left over?

This dilemma led cosmologists to search for some sort of mechanism for leaving the universe as we see it today; and they now believe that they know how this happened. The laws of physics at super-high levels of energy and temperature did not apply to the same extent or equally to particles and antiparticles. Nature has a definite preference for matter.

Normally, under the energy and temperature levels prevailing in the present universe, nature follows symmetry. There are three basic symmetries in particle physics: C, P, and T. The C (charge conjugation) symmetry specifies that the laws of physics are the same for particles and antiparticles. Symmetry P (parity) deals with the mirror image of particles, right- and left-spinning electrons, for instance. Finally, the T (time reversal) symmetry tells us that the laws are the same in the forward and backward direction of time.

There are, however, exceptions to the C, P, and T symmetries in the present universe. In 1956 Chen-Ning Yang and Tsung-Dao Lee discovered that P was not conserved in the weak interactions. Physicists soon discovered that if they coupled parity (P) with charge conjugation (C), a process in which the particles are replaced by their antiparticles (and vice versa), conservation was satisfied. This new process, referred to as CP conservation, was considered to be universally valid. Then, in 1964, James Cronin and Val Fitch discovered that CP was also not conserved. While studying an exotic subnuclear particle called neutral K mesons (K^0), they found that it decayed at a different rate than its antiparticle. The effect was exceedingly small but of momentous significance. The assumption that all physical processes are symmetric between matter and antimatter was shown to be false. There *is* a tiny asymmetry.

It turned out that the CP nonconservation was the key to understanding how the early symmetric universe evolved into one containing matter only. Within a year after the discovery Andrei Sakharov of Russia showed how to use CP violation to explain our present universe. Sakharov's work was so far ahead of its time, generally ignored outside of Russia for over 10 years. Finally, with the advent of grand unified theories, it was rediscovered.

Sakharov soon realized that violation of CP (and C) conservation was not enough. Nonconservation of baryon number was also needed.

What is baryon conservation? Heavy particles such as protons and neutrons are known as baryons. Physicists have found it convenient to label baryons and other particles with various quantum numbers; in practice these numbers are little more than a bookkeeping device. One quantum number is called baryon number (B). Protons, neutrons, and all other baryons are given the baryon number $B = 1$. Antiprotons, antineutrons, and other antibaryons are given the baryon number $B = -1$. All other particles are assigned $B = 0$.

Baryon conservation means that in any interaction the total baryon number B remains constant. Whatever B is before the interaction, it must be the same after. And for years scientists were convinced that B was conserved. There was, in fact, a strong reason for their belief. If it was not conserved the proton would be unstable and would decay. Everyone was confident that this did not happen. If it did decay, and had a lifetime of less than 10^{16} years, physicists would be able to detect radiation coming from our bodies.

However, Sakharov showed that the nonconservation of baryon number would be needed to leave the universe with its preponderance of matter. Furthermore, he specified that the universe must go from a state of equilibrium to one of nonequilibrium. With these two conditions and CP violation, he said the universe could end up the way we see it today. Furthermore, Sakharov calculated the expected lifetime of a proton and got a large but finite number, of about 10^{31} years. How could we ever measure it? If we assemble, say, 10^{34} protons, one of them should decay every few days. And, as it turns out, 10^{34} protons is not an overwhelming number; they could easily be housed in a small building. Another advantage of such an experiment is that protons of one material are the same as protons of another. We can therefore use relatively cheap materials such as water or iron. Using such materials, several experiments have been set up: one in an old gold mine in India, another in a tunnel under Mont Blanc on the border of Italy and France, yet another in an old salt mine in Ohio, and several others at still other locations. So far, unfortunately, no one has caught a proton in the act of decaying. But most physicists are convinced they will eventually detect the decay.

In summary, the essentials of the problem were laid out by Andrei Sakharov– namely skewing the universe toward matter required two things: some means of converting matter to antimatter and vice versa (baryon-number-conservation violation) and some matter-antimatter asymmetry that would make this process favor the direction of matter (CP violation). But having proposed these conditions, he conceded that there were few clues (at the time, around 1967) as to how these conditions might have been met.

As mentioned above, Val Fitch and James Cronin observed CP violation in the decay of kaon (K^0). K^0 decayed at a different rate than its antiparticle. But it was too weak by 10 orders of magnitude to meet Sakhorov's conditions. As for baryon number violating processes, Gerared 't Hooft discovered in 1975 that the standard model of particle physics predicted that matter should be able to tunnel into antimatter, and vice versa, in much the same way as an electron can quantum tunnel through an energy barrier. The so-called 't Hooft effect was extremely small, allowing no more than about one particle in 10^{120} to make the switch. Sakhorov's conditions required about one in a billion. Then, in 1985, Russian physicist Mikhail Shaposhnikov and his collaborators argued that the 't Hooft effect might supply the requisite baryon number violation after all. In spite of its rarity at familiar energies, they conjectured that at the very high energies that prevailed in the early universe the effect would be vastly amplified. This still left us in need of a source of CP violation.

After years of efforts, experimental and theoretical physicists have found a natural way for CP violation within the standard model: The B mesons might reveal considerably more about CP violation than kaons possibly could. B mesons are similar to kaons, but have a strange quark replaced by the much more massive bottom quark. Calculations indicate that CP violation might be 100 times greater for B mesons than for kaons. A careful study of CP violation in the decay of the meson and the anti-B meson would be extremely interesting.

Much higher intensities of B mesons are required. Most electron accelerators like those at Stanford and Cornell in the United States, KEK in Japan, and DESY in Germany, will produce the upsilon particle in head-on collisions of electrons and positrons at the resonance energy of 10.58 GeV, which in turn decays into a B meson and anti-B meson. The proton accelerators at Fermilab and CERN have also been used to study B decay asymmetries at much higher energies. Intriguing indications of CP violation in B meson has turned up at Fermilab. The next few years promise to be a rich harvest of B meson experiments.

With one of Sakharov's two elements—baryon-number violation—in hand and the other—adequate CP-symmetry violation—a good bet for the near future, American physicists Dine and McLerran went on to build complete scenarios for the origin of the matter asymmetry. Both scenarios rely on a version of the inflationary model, in which the newborn universe goes through an episode of sudden inflation and then experiences a phase transition analogous to the boiling of a liquid. As steam appears as bubbles in water, a new phase of the universe emerges as expanding bubbles. Outside the bubbles is a hot soup of massless particles in which the direction of time is ill defined, while inside the bubbles are matter and time much as we know them.

The challenge was to find a point in this bubbling universe in which both the amplified 't Hooft effect and the outsized CP violation were holding sway.

Inside the bubbles wouldn't do, because the energy there was too low to enlarge the 't Hooft effect. Outside the bubbles wouldn't work either, because time was ill-defined time there, which enabled CP violation to work in both directions, canceling out any matter-antimatter imbalance. The only place where CP violation and the 't Hooft effect would have overlapped was in the walls of the bubbles. Eureka, went the thinking: All the matter in our universe is simply a relic of processes in those short-lived bubble walls.

Dine's and McLerran's groups calculated the amount of excess matter their scenarios can generate. They found that the results jibed with estimates of the matter-antimatter imbalance that must have prevailed in the early universe.

It is a very clever mechanism, but we are not sure there are not other scenarios that might also work. Certainly the case is not closed. After all, half of the scenario—the CP-violation part—is still waiting for the verdict from B factory at SLA. But the first finding from SLC is not the kind Dine and McLerran were hoping for—still short by a factor of a billion.

11.6 The Dark Matter Problem

The distribution of matter in the universe today is quite lumpy. Stars are grouped together in galaxies, galaxies into clusters, and clusters into superclusters that stretch across 50 Mpc. The distribution of matter during the early universe must not have been perfectly uniform. If it had been, it would still have to be absolutely uniform today; there would now be only a few atoms per cubic meter of space, with no stars and no galaxies. Consequently, there must have been a slight lumpiness, or density fluctuations, in the distribution of matter in the early universe. Through the action of gravity, these fluctuations eventually grew to become the galaxies and clusters of galaxies that we see today throughout the universe.

Our understanding of how gravity can amplify density fluctuations dates back to 1902, when British physicist James Jeans solved the problem of how the region of higher density gravitationally attract nearby matter and gas mass. As this happens, however, the pressure of the gas inside these regions will also increase, which can make these regions expand and disperse. The question then becomes: Under what conditions does gravity overwhelm gas pressure so that a permanent object can form?

Jeans proved that an object will grow from a density fluctuation, provided that the fluctuation extends over a distance that exceeds the so-called Jeans length L_J:

$$L_J = \sqrt{\pi kT/(mG\rho)}$$

where k = Boltzmann constant = 1.38×10^{-23} J/K, T = temperature of the gas (in Kelvin), m = mass of a single particle in the gas, G (universal constant of gravitation) = 6.67×10^{-11} N · m^2/kg^2, and ρ_m = average density of matter in the gas.

Density fluctuations that extend across a distance larger than the Jeans length tend to grow, while fluctuations smaller than L_J tend to disappear.

We can apply the Jeans formula to the conditions that prevailed during the era of recombination, when $T = 3000\,\mathrm{K}$ and $\rho_m = 10^{-18}\,\mathrm{kg/m^3}$. Taking m to be the mass of the hydrogen atom ($1.67 \times 10^{-27}\,\mathrm{kg}$), we find that $L_J = 100$ light-years, the diameter of a typical globular cluster. Moreover, the mass contained in a cube whose sides are $L_J (= \rho_m \times L_J^3)$ is about $5 \times 10^5 M_\odot$, equal to the mass of a typical globular cluster. Globular clusters contain the most ancient stars we can find in the sky. For these reasons, Robert Dicke and P. J. Peebles proposed that globular clusters were among the first objects to form after matter and radiation decoupled from each other. Objects the size of globular clusters may have merged to form still larger collections of matter; over time, such mergers may have led to the population of galaxies we see today.

If large structure did indeed develop from density fluctuations, how did this development take place? Observations of nonuniformity in the background radiation made with COBE show fluctuations of about 1 part in 100,000. Such small fluctuations at the time of recombination are far too small to explain the large structures in the universe we see today. Gravity is not strong enough to grow galaxies from such small fluctuations within a reasonable time. Model calculations indicate that for density fluctuations in the very early universe to collapse to form today's galaxies, those fluctuations must be at least 0.2% greater than the average density. If normal matter in the very early universe had been clumped at this scale, the variations in the cosmic background radiation today would be at least 30 times larger than what is observed. How do we solve this discrepancy? Once again, dark matter is at the recourse! Physicists and cosmologists believe that 90% of the mass in the universe is in the form of dark matter. They cannot baryonic matter, made up of neutrons and protons. Otherwise the density of neutrons and protons in the early universe would be very much higher, and hence the abundances of light elements in the universe would be very much different from what we actually observe.

How could dark matter solve the structure problem of the universe? Although the radiation interacting with protons and electrons in the plasma of the early universe may prevent the clumping of ordinary matter until after atoms are formed, there is no reason why the same should be true of dark matter. Suppose (for the sake of argument for the moment) that we had a candidate for dark matter that stopped interacting with radiation very early in the Big Bang–during the first second, for example. This situation could arise if the interaction of the dark matter particles with radiation depended on the energy of collisions between the two and hence became small once the temperature fell below a certain level. In such case, the dark matter could start to come together into clumps under the influence of gravity long before the formation of (ordinary) atoms. If this happened, then when normal matter finally formed, it would find itself in a universe in which enormous concentrations of mass (of dark matter) already existed. Bits of ordinary matter would be strongly attracted to the places where dark matter had already congregated and would move quickly to those spots, and thus galaxies and other structures could form very quickly after radiation decouples.

At this point we have a notion that dark matter might work. To go from the notion to a theory, we have to answer two important and difficult questions: (1) What is the dark matter? and (2) How does the dark matter explain structure? The nature of the unseen dark matter is not known yet. This does not prevent physicists from hypothesizing different types of dark matter in the hope of explaining the large structures that we see. Dark matter may be baryonic or nonbaryonic. Gas or dust clouds are the first things that come to astronomers' minds as candidates for baryonic matter, but we find them to be insufficient. An exotic candidate would be black holes because they are not luminous, and if they are big enough, they have long lifetimes. They are believed to sit at the centers of galaxies and have masses exceeding $100M_\odot$. But this is not a solution to the galactic rotation curves because dark matter is needed in the haloes, not in the centers of galaxies.

More serious candidates of baryonic matter are brown dwarfs, stars with masses less than $0.08 \, M_\odot$. They also go under the acronym MACHO for Massive Compact Halo Objects. They lack sufficient pressure to start hydrogen burning, and so their only source of luminous energy is the gravitational energy lost during slow contraction. Such objects would clearly be very difficult to see. But if a MACHO passes exactly in front of a distant star, the MACHO would act as a gravitational lens—the light from the star would bend around the massive object. The intensity of starlight would then be temporarily amplified, and this lensing effect by MACHO could be detected. The difficulty is that we have to monitor millions of stars for one positive piece of evidence. A few lensing effects by MACHOs have been discovered in the space between Earth and the Large Magellanic Cloud.

Recently, two teams of astronomers reported that these massive compact halo objects might be nothing more than elderly white dwarfs. Rodrigo A. Ibata of the European Southern Observatory in Garching, Germany, and Harvey B. Richer of the University of British Columbia in Vancouver used Hubble to reexamine the Hubble Deep Field North, two years after the telescope first imaged this region of the sky. By comparing the two image sets they picked out five extremely faint objects that had moved slightly. Remote galaxies do not move perceptibly across the sky, so the objects must reside in or near the Milky Way. Their particular motion, brightness, and bluish color suggest they are faint white dwarfs a few thousand light-years from Earth.

On the other hand, Rene A. Mendez of the Cerro Tololo Inter-American Observatory near La Serena, Chile, and Dante Minniti of the Pontificia Universidad Catolica de Chile in Santiago analyzed single images of the Hubble Deep Fields, North and South. They found 15 pointlike sources of light whose bluish color is indicative of old white dwarfs. These objects are likely to lie in the halo less than 6,000 light-years from Earth. The team couldn't determine whether the 15 objects have detectable motion, but tests show that they aren't remote galaxies. A preliminary analysis by Ibata's team suggests that these 15 objects do not include the five found by comparing old and new images.

If the findings hold up, they could revolutionize the way astronomers think about the Milky Way and perhaps the structure of all galaxies because formation of such objects would have thrown into interstellar space far more carbon, oxygen, and nitrogen than observations show. In addition, the appearance of galaxies today does not indicate that they once had enough sunlike stars to form a large population of halo white dwarfs.

The findings may be intriguing, but the results won't solve the mystery of dark matter throughout the universe. The Big Bang theory predicts that most dark matter must be of some exotic form, not baryonic.

The nonbaryonic dark matter can be classified in two groups: hot dark matter, consisting of light particles that were still relativistic at the time of their decoupling, and cold dark matter particles that are either quite heavy and that therefore decoupled early, or superlight particles with superweak interactions that were never in thermal equilibrium. Neutrinos are an example of hot dark matter, and examples of cold dark matter include WIMPs (weakly interacting massive particles) as well as other even more speculative exotic particles. These two types of dark matter lead to quite different kinds of structure in the present-day universe. By performing computer simulations of model universes dominated by hot and by cold dark matter and comparing the results with observations of the real universe, cosmologists try to determine which, if either, of the two alternatives can account for the large-scale structure we see around us.

Many researchers suspect that neutrinos have a small mass; this would make them leading candidates for hot dark matter particles. The neutrino has generally been thought to be massless because of a conservation law known as the conservation of lepton number. The neutrino that spins to the left as it moves (i.e., in the direction of the curled fingers of your left hand if your thumb points in the direction of the neutrino motion) is assigned a lepton number of $+1$, and the antineutrino that spins to the right has a lepton number of -1. The electron (both left- and right-handed) is assigned a lepton number of $+1$, and the positron has a lepton number of -1. Conservation of lepton number means that the total lepton number of any system cannot change. Now, if the neutrino had a mass, then it would always be traveling at a speed less than that of light, and the distinction between left- and right-handed spin would lose its absolute significance. By traveling sufficiently fast past a neutrino one could reverse the apparent direction of its motion but not its spin, thus converting a left-handed neutrino into a right-handed antineutrino by a mere change of point of view. If lepton number is conserved this would be supposed to be impossible, so to avoid a contradiction we would have to suppose that the neutrino is massless, so that no observer can ever travel faster than it does. (This argument does not apply to the electron, because both electron and its antiparticle come with both spins, left and right.)

There are indications from laboratory experiments and astronomical events (Supernova 1987A) that neutrinos may have masses of 10 to 40 eV. (For comparison, the electron mass is 0.511 MeV.) This would be enormously important, because there are expected to be about as many neutrinos and antineutrinos left over from the early universe as there are photons in the microwave background radiation, or about 10,000 million neutrinos and antineutrinos for each proton or neutron. A neutrino mass of 10 or more electron volts would therefore mean that it is neutrinos rather than nuclear particles that provide most of the mass density of the universe. This will certainly close the universe. Also, massive neutrinos are not subject to the nongravitational forces that allow nuclear particles and electrons to collapse into the central parts of galaxies, so they are good candidates for the mysterious dark matter in the outer reaches of galaxies and in clusters of galaxies.

Two factors affect the interaction of neutrinos with (ordinary) matter: the density of matter (which tells us how often neutrinos come near other particles) and the probability that a neutrino coming near another particle will actually interact with it. After the universe was one second old, this combined probability was low enough for neutrinos to decouple from matter. Therefore, the neutrinos expanded and cooled on their own, analogous to the cosmic microwave background radiation.

Streaming neutrinos tended to break up small mass concentrations but would leave large ones more or less untouched. Such a selective demolition of certain mass concentrations in the early universe would occur long before radiation decoupled and gravitational collapse. This would mean that the neutrinos had destroyed every nucleus around which galaxies of less than a certain size could condense. Simulations of a universe filled with hot dark matter confirm this: large structures, such as superclusters and voids, can form fairly naturally; however, computer models cannot account for the existence of structure on smaller scales. Small amounts of hot material tend to disperse, not to clump together. Attempts to produce galaxies and clusters by other means after the formation of larger objects have been only partly successful, so most cosmologists believe that models based purely on hot dark matter are unable to explain the observed structure of the universe.

Cold dark matter avoids this difficulty; small groups of mass come together first, and these small aggregates gather to form larger structures. The cold dark matter predicts that galaxies would be created in a rather restricted mass range: from about one-thousandth (10^{-3}) to about ten thousand (10^4) times as massive as the Milky Way—none bigger or smaller. It is interesting to note that almost all known galaxies have masses within this range. However, the results of recent redshift surveys and the discovery of the voids and filaments created serious problems for cold dark matter as the ultimate constituent of the structure of the universe. But clever ideas do not die easily. Marc Davis and his group at UC Berkeley argued that when radiation decoupled, luminous matter would not be scattered uniformly in space, but would tend to gather where large amounts of dark matter already existed. As we look out at the universe we are not seeing the regions where all the dark matter is, but only the places where it has pulled in enough luminous matter to create a galaxy or a galactic cluster. The Berkeley group reasoned that it is very possible that dark matter is spread much more uniformly than luminous matter, so that the voids we see may actually have dark matter in them. Using this line of reasoning, Marc Davis and his Berkeley group have produced plots of galaxy distribution that look very much like what is actually observed. They still have trouble producing large voids with sharp edges.

Cosmologists first believed that with some fine tuning, these models could also be made to produce large-scale structures comparable to what is actually observed. They are now not that certain, because the ripples seen by George Smoot and his COBE team in 1992, taken in conjunction with standard cold dark matter, imply too little structure on large scales (superclusters, voids, and so on). At the time of writing, the status of cold dark matter models as the proper description of invisible matter in the universe is still not quite certain.

Some physicists are exploring a new scenario in which the universe contains a nearly even blend of hot and cold dark matter. They search for ways to create two kinds of dark matter by a single mechanism. They propose that the universe initially contained a population of massive neutrinos; these neutrinos could have decayed in a way to stimulate the formation of slow-moving cold dark matter. This mechanism is called neutrino lasing, analogous to the stimulated creation of photons of light in a conventional laser. The heavy neutrinos themselves decay into lighter, high-speed particles that constitute the hot dark matter. In this way, a single, fairly elegant set of events can account for the existence of two separate components of dark matter. Neutrino lasing occurs at such high energies that we cannot devise a laboratory test for it.

At this point, WIMPs are the leading suspect for cold dark matter. Physicists theorize that these tiny, weighty particles (estimated to be 50 times heavier than a proton), which originated during the Big Bang, became nonrelativistic much earlier than leptons and decoupled from the hot plasma. For example, the supersymmetric models contain at least three such particles (photino, zino, and gaugino). They only interact weakly with the protons and neutrons of the visible universe. If real, 10 trillion WIMPs may be zipping through every 2 pounds of matter here on Earth every second. A dozen experiments worldwide are based on the assumption that occasionally a WIMP might smack into normal matter. The challenge has been to differentiate them from other particles that zip through the cosmos.

Recently physicists working on DAMA (the Italian dark matter experiment) announced that they possibly found the elusive particles. The DAMA is a mile underground and uses ultracold detectors that emit flashes of light whenever a particle collides with sodium iodide atoms. The DAMA experiment could differentiate possible WIMPs from charged particles; it could not distinguish the elusive mystery matter from ordinary neutrons. Although the experiment is a mile underground, it is shielded from most but not all stray neutrons.

A more discriminating detector cooled to near absolute zero and buried 30 feet beneath Stanford University registered hits like the DAMA; detailed analysis showed the events were most likely caused by neutrons. In addition to registering hits, the Stanford team also makes two specific measurements: the amount of heat released and the amount of electricity that is discharged. These two different kinds of information let the American team see a much clearer picture of what is causing the event. The Stanford experiment soon will be moved to an abandoned iron mine in northern Minnesota, where it will be shielded by 4,300 feet of rock. Sensitivity is expected to increase by a factor of 100 when the $12 million, six-year project gets under way. At least five other similar cryogenic experiments are being built or are planned around the world. Other researchers are focusing on creating the particles with high-speed accelerators.

The discovery of WIMPs not only helps physicists determine the mass of the universe; it also can validate a popular and elegant theory that predicts a yet-to-be-found partner for every known particle.

If cosmic strings do exist, some physicists speculate that they might be a candidate for dark matter in the universe, but they do not fit into the hot and cold scheme of classification; in fact, they are not made up of particles at all. What are cosmic strings? To answer this question let us revisit Guth's inflationary theory. The inflation occurred as the universe went from a high-energy symmetric state of a false vacuum to a low-energy symmetry-broken state of a true vacuum. Now a question arises naturally: Could defects exist in the early breakdown of symmetry of the forces? A. Vilenkin of Tufts University believed so and raised this question in 1977. These defects are known as cosmic strings, and may be compared with the breakdown of symmetry when ice forms from water. Slight impurities cause the production of defects in the crystallized ice. Likewise, any slight randomness in the early, unified universe causes imperfections that show up as defects later on.

Cosmic strings are bits of isolated space containing very high energies. Perhaps cosmic strings might be better described as pure bits of unified force fields in which the strong, electromagnetic, and weak forces remain unified. They are incredibly narrow, only about 10^{-10} km wide, and may form long loops billions of kilometers long; thus, the term *strings*. As they are defects or distortions of space, they produce an effect of great mass concentration. J. Ostriker and Ed Witten have found an even more fascinating property of the strings; they may not just attract matter around them, they may also repel it. This is caused by their strong radiating tendency. This radiation can be intense enough to push away matter near the string, effectively clearing out a channel of space for billions of kilometers.

What evidence do we have of cosmic strings? Many voids (regions free of galaxies) had been found in recent years. A curious aspect of these voids was that clusters of galaxies bordered the edge of the void in a way that showed a gigantic filament-type structure. The voids and filaments don't make sense with the standard Big Bang model. Many physicists and cosmologists suspect that cosmic strings could be responsible for these voids and filaments.

Cosmic strings are very appealing as an explanation for voids. At one time cosmic strings were considered as good candidates of dark matter, but now few physicists believe that they are a vital component of dark matter.

11.7 The Primordial Magnetic Fields

In Big Bang models the primordial magnetic fields is neglected, mostly for simplicity. But, according to British physicist Christos Tsagas's study, magnetic fields could have an interesting cosmological effect. As we have learned in the first three chapters, Einstein's General Theory of Relativity is essentially a description of the geometry of space and time. Like very elastic rubber bands under tension, magnetic field lines try to remain as straight as possible. Magnetic fields transmit that tension to space-time, making nearby space seem like a rubber sheet that has been stretched a little bit tighter. Such a region becomes stiffer and flattens out somewhat. This effect can be significant.

If the Big Bang created a primordial magnetic field, the extra stiffness of space-time would have resisted the rapid inflation that was believed to have occurred around 10^{-35} s after the Big Bang. It also would have ironed out the entire universe, making the background cosmology more like a flat cosmological model. That might help explain why the cosmos does not appear to have any curvature. Stiffer space-time might also damp gravitational waves and make them harder to detect than physicists at observatories such as LIGO and TAMA have been counting on. Black hole theorists who deal with sharply curved space near strong magnetic fields might need to revise some pet notions as well. Cosmologists need to rethink the role of magnetic fields in shaping the cosmos.

11.8 Problems

11.1. In a radiation-dominated universe, temperature and time are related by (11.11); and in a matter-dominated universe, they are related by the following equation:

$$T(K) \cong 1.9 \times 10^{12} t^{-3/2}(s).$$

Derive these two equations by applying the conservation of energy (the first law of thermodynamics) for a sample volume V: $dE + PdV = 0$, where P is the pressure and E is the matter-energy density in $V (E = \rho c^2)$.
Hint: for the radiation-dominated case, $P = (1/3)\rho c^2$, and $P = 0$ for the matter-dominated case (noninteracting matter). Also consider a flat geometry (the transition case).

11.2. As shown in the text, the ratio between the temperature at the decoupling era and the present temperature is given by

$$T_d/T_0 = 3000K/3K = 1000.$$

Show that the universe was about $1/1000^{\text{th}}$ its present size at the time of decoupling.

11.3. Sakharov showed that there is an upper boundary

$$T < T_{\text{max}} = (\alpha/k)(\hbar c^5/G)^{1/2} \sim 10^{32} K$$

on the temperature of blackbody radiation. In the equation α, k, $\hbar$, c, and G denote, respectively, a constant factor near unity, the Boltzmann constant, the Planck constant, the speed of light, and the gravitational constant. Sakharov's deduction starts from the thermodynamic properties of the hot matter in the isotropic universe within the frame of gravitational perturbation. Try to give a simpler deduction by considering a spherical box of radius R filled with blackbody radiation of temperature T and energy density $\rho = aT^4$ (a is the Stefan constant), and assume that the gravitational field of the box can be described by a simple Schwarzschild metric. (Gravitational constraints on blackbody radiation, Corrado Massa, *American Journal of Physics.* 54: 754, 1988.)

References

Barrow J, Silk J (1993) *The Left Hand of Creation: Origin and Evolution of the Universe.* (Oxford University Press, UK)

Liddle A (1999) *An Introduction to Modern Cosmology.* (John Wiley & Sons Ltd, Chichester, England)

Sakharov AD (1967) Violation of CP invariance, C and baryon asymmetries of the universe. JETP Lett 5:32–5

Weinberg S (1988) *The First Three Minutes.* (Basic Books, Inc., New York)

Appendix A
Classical Mechanics

Classical mechanics studies the motion of physical bodies at the macroscopic level. Newton and Galileo first laid its foundations in the 17th century. The essential physics of the mechanics of Newton and Galileo, known as Newtonian mechanics, is contained in the three laws of motion. Classical mechanics has since been reformulated in a few different forms: the Lagrange, the Hamilton, and the Hamilton-Jacobi formalisms. They are alternatives, but equivalent to the Newtonian mechanics. We will review Newtonian mechanics first and then Lagrangian dynamics.

A.1 Newtonian Mechanics

A.1.1 The Three Laws of Motion

The foundations of Newtonian mechanics are the three laws of motion that were postulated by Isaac Newton, resulting from a combination of experimental evidence and a great deal of intuition. The first law states:

A free particle continues in its state of rest or of uniform motion until an external force acts upon it.

Newton made the first law more precise by introducing the concepts of "quantity of motion" and "amount of matter" that we now call momentum, and mass of the particle, respectively. The momentum $\vec{p}$ of a particle is related to its velocity $\vec{v}$ and mass m by the relation

$$\vec{p} = m\vec{v}$$

The first law may now be expressed mathematically as

$$\vec{p} = m\vec{v} = \text{constant}$$

provided that there is no external force acting on the particle. Thus, the first law is a law of conservation of momentum.

The second law gives a specific way of determining how the motion of a particle is changed when a force acts it on. It may stated as

The time rate of change of momentum of the particle is equal to the external applied force:

$$\vec{F} = d\vec{p}/dt.$$

The second law can also be written as, when mass m is constant, as

$$\vec{F} = md\vec{v}/dt,$$

which says that the force acting on a particle (of constant mass) is equal to the product of its acceleration and its mass, a familiar result from basic physics.

Mass that appears in Newton's laws of motion is called the inertial mass of the particle, as it is indicative of the resistance offered by the particle to a change of its velocity.

The third law states that

The force of action and reaction are equal in magnitude but opposite in direction.

Physical laws are usually stated relative to some reference frame. Although reference frame can be chosen arbitrarily in an infinite number of ways, the description of motion in different frames will, in general, be different. There are frames of reference relative to which all bodies that do not interact with other bodies move uniformly in a straight line. Frames of reference satisfying this condition are called *inertial frames of reference.* It is evident that the three laws of motion apply in inertial frames. In fact, some scientists suggest that Newton singled out the first law just for the purpose of defining inertial frames of reference. But the true significance of Newton's first law is that, in contrast to the view held by his science contemporaries or those before him, the state of a body at rest is equal to that of a body in uniform motion (with a constant velocity in a straight line).

Then how is it possible to determine whether or not a given frame of reference is an inertial frame? The answer is not quite as trivial as it might seem at first sight, for in order to eliminate all forces on a body, it would be necessary to isolate the body completely. In principle this is impossible because, unless the body were removed infinitely far away from all other matter, there are at least some gravitational forces acting on it. Therefore, in practice we merely specify an approximate inertial frame in accordance with the needs of the problem under investigation. For example, for elementary applications in the laboratory, a frame attached to Earth usually suffices. This frame is, of course, an approximate inertial frame, owing to the daily rotation of Earth on its axis and its revolution around the sun. It is also a basic assumption of classical mechanics that space is continuous and the geometric of space is Euclidean, and that time is absolute. The assumptions of absolute time and of the geometry of space have been modified by Einstein's theory of relativity, which we will discuss in the following sections.

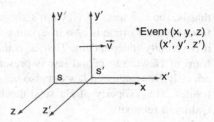

Fig. A.1 Coordinate frames S and S′ moving with constant relative velocity.

A.1.2 The Galilean Transformation

Any frame moving at constant velocity with respect to an inertial frame is also an inertial frame. To show this, let us consider two frames, S and S′, which coincide at time $t = t' = 0$, as shown in Figure A.1. S′ moves with a constant velocity V relative to S. The corresponding axes of S and S′ remain parallel throughout the motion. It is assumed that the same units of distance and time are adopted in both frames. An inspection of Figure A.1 gives

$$\vec{r}' = \vec{r} - \vec{V}t \quad \text{and} \quad t' = t \tag{A1-1}$$

The relations (A1-1) are called the *Galilean transformations*. The second relation expresses the universality, or absoluteness, of time. We shall see later that the Galilean transformations are only valid for velocities that are small compared with that of light.

Differentiating (A1-1) once, we obtain

$$\dot{\vec{r}}' = \dot{\vec{r}} - \vec{V}, \tag{A1-2}$$

which says that if the velocity of a particle in frame S is a constant, then its velocity in frame S′ is also a constant. Differentiating the velocity relation (A1-2) once with respect to time, we obtain

$$\ddot{\vec{r}}' = \ddot{\vec{r}}, \tag{A1-3}$$

which indicates that the acceleration of the particle is also a Galilean invariant.

If frame S in Figure A.1 is inertial, so is S′, since the linear equations of motion of *free* particles in frame S are transformed by (A1-1) into similar linear equations in S′. Any frame that moves *uniformly* (i.e., with constant velocity and without rotation) relative to any inertial frame is also itself an inertial frame. And there is an infinity of inertial frames, all connected by Galilean transformations.

A.1.3 Newtonian Relativity and Newton's Absolute Space

The inertial mass m is also a Galilean invariant, so

$$m\ddot{\vec{r}}' = m\ddot{\vec{r}},$$

that is, the $m\vec{a}$ term of Newton's second law is a Galilean invariant. The applied force on the particle is also invariant under Galilean transformations, provided that it is velocity-independent. That is, if the applied force is velocity-independent, the form of Newton's second law is preserved under Galilean transformations. Thus, not only Newton's first law but also his second and third laws are valid in all inertial frames. This property of classical mechanics is often referred to as Newtonian (or Galilean) relativity.

Because of this relativity, the uniform motion of one inertial frame relative to another cannot be detected by internal mechanical experiments of Newton's theory. According to the first law, a particle does not resist uniform motion, of whatever speed, but it does resist any change in its velocity, i.e., acceleration. Newton's second law precisely expresses this, and mass m is a measure of the particle's inertia. Here we should ask, "Acceleration with respect to what?" One may give a simple answer: with respect to any one of the inertial frames. However, this answer is quite unsatisfactory. Why does nature single out such "preferred" frames as standards of acceleration? Inertial frames are unaccelerated and nonrotating, but relative to what? Newton found no answer and postulated instead the existence of an absolute space, which "exists in itself, as if it were a substance, with basic properties and quantities that are not dependent on its relationship to anything else whatsoever (i.e., the matter that is in this space)." Newton's concept of an absolute space has never lacked critics. From Huyghens, Leibniz, and Bishop Berkeley to Ernst Mach and Einstein, these objections have been brought against absolute space:

(1) It is purely *ad hoc* and explains nothing.
(2) How are we to identify which inertial frame is at rest relative to absolute space?
(3) Newton's absolute space is a physical entity; thus it acts on matter (it is the "seat of inertia" resisting acceleration in the absence of forces). But matter does not act on it. As Einstein said: "It conflicts with one's scientific understanding to conceive of a thing (absolute space) that acts, but cannot be acted upon."

Objection (3) is perhaps the most powerful one. It questions not only absolute space but also the set of all inertial frames.

Newton's theory can do without absolute space. Space can be regarded as a concept necessary for the ordering of material objects. From this viewpoint space is just a generalization of the concept of place assigned to matter. Matter comes first and the concept of space is secondary. Empty space has no meaning. The ancient Greeks, led by the great philosopher Aristotle, thought about space in this fashion. If we take this view of space, then the space is not absolute. As we saw above there is an infinity of inertial frames, connected by Galilean transformations. One inertial frame of reference does not in any way differ from the others. A reference frame attached to Earth's surface can be considered as an inertial frame, and a train moving with a constant velocity with respect to Earth is also an inertial frame; the laws of motion would hold inside the train. If a rubber ball on the train bounces straight up and down, it hits the floor twice on the same spot a certain time apart, say one second. However, to someone standing by the rails, the two bounces would take place a certain distance apart. Thus we cannot give an event an absolute position in space. The lack of absolute position means that space is not absolute.

A.1.4 Newton's Law of Gravity

Historically, Newton began with his laws of motion and Kepler's laws of planetary motion to arrive at his law of gravity. Kepler discovered three laws that planets obey as they move around the sun. These laws were based on Tycho Brahe's observational data on Mars' motion around the sun:

The first law: Each planet revolves around the sun in an elliptical orbit, with the sun at one focus of the ellipse.

The second law: The speed of a planet in its orbit varies in such a way that the radius connecting the planet and the sun sweeps over equal areas in equal time intervals. Figure A.2 illustrates the second law. Each planet moves fastest when it is nearest the sun, and slowest when it is farthest away.

The third law: The ratio between the square of a planet's period of revolution T and the cube of the major axis a of its orbit has the same value for all the planets,

$$T^2/a^3 = K$$

where K is a constant, the same for all the planets.

For the derivation of the law of gravity from these laws, see my text, *Classical Mechanics*, or any other textbook on classical mechanics. Here is a simplified version: As the orbits of the planets are very near to being circles, we assume for simplicity that the planets move in circular orbits around the sun. Now, a planet of mass m circulating the sun at a radius r with a velocity v is acted upon by the centripetal force

$$F_c = mv^2/r.$$

The distance a planet travels in one revolution is the circumference of the circle, $2\pi r$. Thus we have

$$2\pi r = vT, \quad \text{or} \quad T = 2\pi r/v$$

where T is the period of revolution; and Kepler's third law becomes

$$\frac{T^2}{r^3} = \frac{4\pi^2}{rv^2} = K,$$

or

$$v^2 = 4\pi^2/Kr.$$

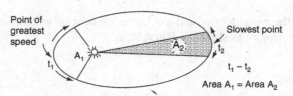

Fig. A.2 Kepler's second law.

Substituting this into F_c we obtain

$$F_c = 4\pi^2 m / K r^2.$$

The centripetal force F_c is provided by the gravitational force of the sun exerted on the planet. Thus,

$$F_g = 4\pi^2 m / K r^2.$$

Now, Newton's third law requires that the force a planet exerts on the sun be equal in magnitude to the force the sun exerts on the planet. This means that F_g must be proportional to both m and the mass of the sun M, and we may express the constant quantity $4\pi^2/K$ as GM. Accordingly, the gravitational force between the sun and any planet is

$$F_g = GMm / r^2.$$

The quantity G is the gravitational constant, and its value is determined experimentally to be 6.67×10^{-11} N · m^2/kg^2.

This is a remarkable result, providing an astonishingly simple and unified basis for the description of the motion of all planets around the sun: it is the gravitational force of the sun that keeps the planets moving in their orbits. Newton went further and made a gigantic extrapolation from his law of gravity by claiming its universality. This extrapolation was based on the argument that the apple, the moon, the planets, and the sun were made of ordinary matter so that they were in no way special. A similar force might well be expected to act between any two masses m_1 and m_2 separated by a distance r (Fig. A.3)

$$\vec{F} = -\frac{Gm_1 m_2}{r^2}\hat{r}. \tag{A1-4}$$

According to (A1-4), gravitational forces act at a distance and are able to produce their effects through millions of miles of empty space. The law also tacitly implies that the gravitational influence propagates with infinite speed.

In the form of (A1-4) the law strictly applies only to point particles or objects that have spherical symmetry. If one or both of the objects has a certain extension, then we need to make additional assumptions before we can calculate the force. The most common assumption is that the gravitational force is linear. That is, the total gravitational force on a particle due to many other particles is the vector sum of all the individual forces. Thus, for example, if m_1 is a point particle of mass m and m_2 is an extended object that has a continuous distribution of matter, then (A1-4) becomes

$$\vec{F} = -Gm \int_V \frac{\rho(\vec{r}')\hat{r}}{r^2} dv' \tag{A1-4a}$$

where $\rho(\vec{r}')$ is the mass density and dv' is the element of volume at the position defined by the vector $\vec{r}'$, as shown in Figure A.4.

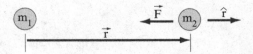

Fig. A.3 Gravitational force on mass m_2 due to mass m_1.

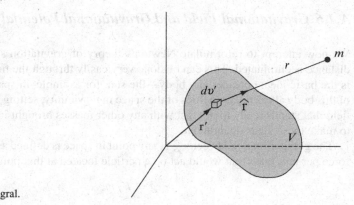

Fig. A.4 Volume integral.

A.1.5 Gravitational Mass and Inertial Mass

The mass of an object plays dual roles. The initial mass m_I, which appears in Newton's laws of motion, determines the acceleration of a body under the action of a given force. The mass entering in Newton's law of gravity determines the gravitational forces between the object and other objects and is known as the object's gravitational mass m_g. The dual role played by mass has an astonishing consequence. If we drop a body near Earth's surface, Newton's second law of motion describes its motion

$$F = m_I a$$

where the force F acting on the object is its weight, the force of gravity of Earth

$$F = G m_g M / r$$

where m_g is the gravitational mass of the falling object, M is the mass of the earth, and r is the distance of the object from Earth's center. Combining these two equations, and if $m_I = m_g$, we then see that the acceleration of an object in any gravitational field is independent of its mass:

$$a = GM/r.$$

Hence, if only gravitational forces act, all bodies similarly projected pursue identical trajectories. Galileo and Newton knew this. Newton conducted experiments to test the equivalence of inertial and gravitational masses. Most recent experiments have found that the two masses are numerically equal to within a few parts in 10^{12}. Einstein was so greatly intrigued by this that he attached a deep significance to it.

The assertion of the equivalence of the inertial mass and gravitational mass is known as the *principle of equivalence*. Pursuit of the consequences of the principle of equivalence led Einstein to formulate his Theory of General Relativity.

A.1.6 Gravitational Field and Gravitational Potential

We now attempt to reformulate Newton's theory of gravitation so that action at a distance is eliminated. This can be done very easily through the field concept. Here is the basic idea: consider any body—the sun, for example–in space. The presence of this body alters the properties of the space in its vicinity, setting up a gravitational field that stands ready to interact with any other masses brought into it. We now try to make these ideas quantitative.

The gravitational field vector $\vec{g}$ at any point in space is defined as the gravitational force per unit mass that would act on a particle located at that point,

$$\vec{g} = \vec{F}/m. \tag{A1-5}$$

Obviously $\vec{g}$ has the dimension of acceleration. Near the surface of Earth, it is called the gravitation acceleration, and its magnitude is about $9.8\,\text{m/sec}^2$.

Now the gravitational force between a pair of masses m and M separated by distance r is

$$\vec{F} = -\frac{GMm}{r^2}\hat{r}$$

where $\hat{r}$ is a unit vector away from M. Therefore the intensity of the gravitational field at distance r from M is

$$\vec{g} = -\frac{GM}{r^2}\hat{r}. \tag{A1-6}$$

If more than one body is present, the gravitational field is the vector sum of the individual fields produced by each body. For a body that consists of a continuous distribution of matter, (A1-6) becomes

$$\vec{g} = -G \int_V \frac{\rho(\vec{r}')\hat{r}}{r^2} dV'. \tag{A1-7}$$

Upon using the identity $\nabla(1/r) = -(1/r^2)\hat{r}$ we find that

$$\vec{g} = G \int_{vol} \nabla(1/r)\rho(\vec{r}')dV'.$$

Since ∇ does not operate on the variable $\vec{r}'$, it can be factored out of the integral and we have

$$\vec{g} = \nabla \int_V \frac{G\rho(\vec{r}')}{r} dV' \tag{A1-8}$$

which can be rewritten as the gradient of a scalar function Φ

$$\vec{g} = -\nabla\Phi$$

with

$$\Phi = -\int_V \frac{G\rho(\vec{r}')}{r} dV'. \tag{A1-9}$$

The quantity Φ has the dimension of energy per unit mass and is called the gravitational potential.

A.1.7 Gravitational Field Equations

The gravitational potential Φ satisfies certain partial differential equations. To find out these equations, we consider a closed surface S enclosing a mass M. The gravitational flux passing through the surface element dS is given by the quantity $\vec{g} \cdot \hat{n} dS$, where $\hat{n}$ is the outward unit vector normal to dS. So the total gravitational flux through S is then given by

$$\int_S \vec{g} \cdot \hat{n} dS = -GM \int_S \frac{\hat{r} \cdot \hat{n}}{r^2} dS. \tag{A1-10}$$

By definition $(\hat{r} \cdot \hat{n}/r^2)dS = \cos\theta \, dS/r^2$ is the element of solid angle $d\Omega$ subtended at M by the element of surface dS (Fig. A.5). And so (A1-10) becomes

$$\int_S \vec{g} \cdot \hat{n} dS = -GM \int d\Omega = -4\pi GM. \tag{A1-11}$$

If the surface S encloses a number of masses M_i, we have

$$\int_S \vec{g} \cdot \hat{n} dS = -4\pi G \sum_i M_i. \tag{A1-12}$$

For a continuous distribution of mass within S, the sum becomes an integral

$$\int_S \vec{g} \cdot \hat{n} dS = -4\pi G \int_V \rho(\vec{r}) dV. \tag{A1-13}$$

Using Gauss' divergence theorem, $\int_S \vec{g} \cdot \hat{n} dS = \int_V \nabla \cdot \vec{g} dV$, (A1-13) becomes

$$\int_V (\nabla \cdot \vec{g} + 4\pi G\rho) dV = 0.$$

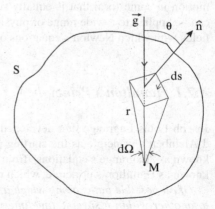

Fig. A.5 The element of solid angle $d\Omega$.

This equation holds true for any volume, and it can be true only if the integrand vanishes:

$$\nabla \cdot \vec{g} = -4\pi G\rho. \tag{A1-14}$$

Since $\vec{g}$ is conservative, a second fundamental differential equation obeyed by $\vec{g}$ is

$$\nabla \times \vec{g} = 0. \tag{A1-15}$$

Now, substituting $\vec{g} = -\nabla\Phi$ into (A1-14) we finally arrive at the result, which is the Poisson's equation:

$$\nabla^2\Phi(\vec{r}) = 4\pi G\rho(\vec{r}) \tag{A1-16}$$

If $\rho = 0$, Poisson's equation reduces to Laplace's equation

$$\nabla^2\Phi(\vec{r}) = 0. \tag{A1-17}$$

Equations (A1-16) and (A1-17) constitute the field equations for the Newtonian theory of gravity.

A.2 Lagrangian Mechanics

As mentioned earlier, classical mechanics can be reformulated in a few different forms, such as the Lagrange, the Hamiltonian, and the Hamilton-Jacobi formalisms. Each is based on the ideas of work or energy, and each is expressed in terms of generalized coordinates. Any convenient set of parameters (quantities) that can be used to specify the state of the system can be taken as the generalized coordinates. Thus, the new generalized coordinates may be any quantities that can be observed to change with the motion of the system, and they need not be geometric quantities. They may be electric charges, for example, in certain circumstances. We shall write the generalized coordinates as q_i, $i = 1, 2, 3, \ldots$ n, where n is the number of degrees of freedom of the system.

In terms of these generalized coordinates, we can write the basic equations of motion in some form that is equally suitable for all coordinate systems, and they can be applied to a wide range of physical phenomena, particularly those involving fields with which Newton's equations of motion are not usually associated.

A.2.1 Hamilton's Principle

Joseph Louis Lagrange first developed the Lagrangian dynamics, and he selected d'Alembert's principle as the starting point and obtained the equations of motion known as "Lagrange's equations" from it. We will start with a variational principle, known as Hamilton's principle, which may be stated as follows:

Of all possible paths along which a dynamical system may move from one point to another within a specific time interval and consistent with any constraints, the actual path followed is that for which the time integral of the difference between the kinetic and potential energies has stationary value.

In terms of the calculus of variation, Hamilton's principle becomes

$$\delta \int_{t_1}^{t_2} (T - V)dt = \delta \int_{t_1}^{t_2} L dt = 0 \tag{A1-18}$$

where L is defined to be the difference between the kinetic and potential energies and is called the Lagrangian of the system. Energy is a scalar quantity, and so the Lagrangian is a scalar function. Hence the Lagrangian must be invariant with respect to coordinate transformations. We are therefore assured that no matter what generalized coordinates are chosen for the description of a system, the Lagrangian will have the same value for a given condition of the system. Although Lagrangian will be expressed by means of different functions, depending on the generalized coordinates used, the value of the Lagrangian is unique for a given condition. Therefore, we can write

$$L = L(q_i, \dot{q}_i, t) = T(q_i, \dot{q}_i, t) - V(q_i, t)$$

where $\dot{q}_i$ is the generalized velocity corresponding to q_i. Hamilton's principle becomes

$$\delta I = \delta \int_{t_1}^{t_2} L(q_i, \dot{q}_i, t)dt \tag{A1-19}$$

where $q_i(t)$, and hence $\dot{q}_i(t)$, is to be varied subject to the end conditions: $\delta q_i(t) = \delta \dot{q}_i(t) = 0$. The symbol "$\delta$" refers to the variation in a quantity between two paths, and "d" as usual refers to a variation along a given path. If we label each possible path by a parameter, δ then stands as shorthand for the parametric procedure outlined below.

We label each possible path by a parameter α in the following way:

$$q_i(t, \alpha) = q_i(t, 0) + \alpha \eta_i(t) \tag{A1-20}$$

where $q_i(t, 0)$ is the actual dynamical path followed by the system (as yet unknown), and $\eta_i(t)$ is a completely arbitrary function of t, which has a continuous first derivative and subject to $\eta_i(t_1) = \eta_i(t_2) = 0$. In terms of the variation symbol, we can write

$$\delta q_i = \frac{\partial q_i}{\partial \alpha} d\alpha = \eta_i d\alpha. \tag{A1-21}$$

This corresponds to a virtual displacement in q_i from the actual dynamical path to a neighboring varied path, as depicted schematically in Figure A.6.

The action integral I is now a function of α only, for any given $\eta_i(t)$:

$$I(\alpha) = \int_{t_1}^{t_2} L\left(q_i(t, \alpha), \dot{q}_i(t, \alpha), t\right)dt \tag{A1-22}$$

Hence, the stationary values of $I(\alpha)$ occur when $\partial I/\partial \alpha = 0$. But by our choice of $q_i(t, 0)$, we know that this occurs when $\alpha = 0$, so that the necessary condition that

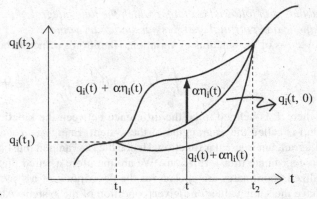

Fig. A.6 The variation of q_i.

the action integral has a stationary value is $\partial I / \partial \alpha = 0$ when $\alpha = 0$. In terms of the variation symbol δ, this necessary condition can be written as

$$\delta I = \left(\frac{\partial I}{\partial \alpha} \right)_{\alpha=0} d\alpha = 0 \tag{A1-23}$$

for arbitrary $\eta_i(t)$ and nonzero α. The subscript $\alpha = 0$ means that we evaluate the derivative $\partial I / \partial \alpha$ at $\alpha = 0$.

A.2.2 Lagrange's Equations of Motion

We now expand the integrand L in (A1-22) in a Taylor's series:

$$I(\alpha) = \int_{t_1}^{t_2} \left[L \left(q_i(t,0), \dot{q}_i(t,0); t \right) + \alpha \eta_i \frac{\partial L}{\partial q_i} + \alpha \dot{\eta}_i \frac{\partial L}{\partial \dot{q}_i} + 0(\alpha^2) \right] dt \tag{A1-24}$$

Since the integration limits t_1 and t_2 are not dependent on α, we can differentiate under the integral sign with respect to α and obtain

$$\frac{\partial I}{\partial \alpha} = \int_{t_1}^{t_2} \left[\frac{\partial L}{\partial q_i} \eta_i + \frac{\partial L}{\partial \dot{q}_i} \dot{\eta}_i + 0(\alpha^2) \right] dt. \tag{A1-25}$$

Dropping terms in α^2, α^3, and integrating by parts the second term we obtain

$$\int_{t_1}^{t_2} \frac{\partial L}{\partial \dot{q}_i} \dot{\eta}_i dt = \left| \frac{\partial L}{\partial \dot{q}_i} \eta_i \right|_{t_1}^{t_2} - \int_{t_1}^{t_2} \eta(t) \frac{d}{dt} \left(\frac{\partial L}{\partial \dot{q}_i} \right) dt \tag{A1-26}$$

The first term on the right is zero because $\eta_i(t_1) = \eta_i(t_2) = 0$. Substituting (A1-26) into (A1-25) we obtain

$$\left(\frac{\partial I}{\partial \alpha} \right)_{\alpha=0} d\alpha = \int_{t_1}^{t_2} \left[\frac{\partial L}{\partial q_i} - \frac{d}{dt} \left(\frac{\partial L}{\partial \dot{q}_i} \right) \right]_{\alpha=0} \eta_i(t) d\alpha dt = 0.$$

To obtain the stationary condition in terms of the variation symbol δ we multiply both sides by $d\alpha$, resulting in

$$\left(\frac{\partial I}{\partial \alpha}\right)_{\alpha=0} = \int_{t_1}^{t_2} \left[\frac{\partial L}{\partial q_i} - \frac{d}{dt}\left(\frac{\partial L}{\partial \dot{q}_i}\right)\right]_{\alpha=0} \eta_i(t) dt$$

or, in terms of the variation symbol δ:

$$\delta I = \int_{t_1}^{t_2} \left[\frac{\partial L}{\partial q_i} - \frac{d}{dt}\left(\frac{\partial L}{\partial \dot{q}_i}\right)\right]_{\alpha=0} \delta q_i(t) dt = 0. \qquad (A1\text{-}27)$$

Since $\delta q_i(t)$ is arbitrary except when $\delta q_i(t_1) = \delta q_i(t_2) = 0$, it follows that a necessary condition for $\delta I = 0$ is that the square bracket vanishes, yielding Lagrange's equations of motion:

$$\frac{d}{dt}\left(\frac{\partial L}{\partial \dot{q}_i}\right) - \frac{\partial L}{\partial q_i} = 0, \quad i = 1, 2, 3, \ldots n \qquad (A1\text{-}28)$$

This is also a sufficient condition for a stationary value of the action integral I. This is from the fact that (A1-28) implies that the integral in (A1-27) vanishes and result in the variation δI being zero.

A.3 Problems

A.1. Under the influence of a force field a particle of mass m moves in the xy-plane and its position vector is given by $\vec{r} = a\cos\omega t\,\hat{i} + b\sin\omega t\,\hat{j}$, where a, b, and ω are positive constants and a > b. Show that

(a) the particle moves in an ellipse;
(b) the force acting on the particle is always directed toward the origin; and
(c) $\vec{r} \times \vec{p} = mab\omega\hat{k}$, and $\vec{r} \cdot \vec{p} = \frac{1}{2}m(b^2 - a^2)\sin 2\omega t$

where p is the momentum of the particle.

A.2. Two astronauts of masses M_A and M_B, initially at rest in free space, pull on either end of a rope. The maximum force with which astronaut A can pull, F_A, is larger than the maximum force with which astronaut B can pull, F_B. Find their motion if each pulls on the rope as hard as possible.

A.3. Show that the electromagnetic wave equation

$$\frac{\partial^2 \phi}{\partial x^2} + \frac{\partial^2 \phi}{\partial y^2} + \frac{\partial^2 \phi}{\partial z^2} - \frac{1}{c^2}\frac{\partial^2 \phi}{\partial t^2} = 0$$

is not invariant under the Galilean transformations.

A.4.(a) Find the force of attraction of a thin spherical shell of radius a on a particle P of mass m at a distance $r > a$ from its center.

(b) Prove that the force of attraction is the same as if all the mass of the spherical shell were concentrated at its center.

A.5. Derive the result of Problem A.4a by first finding the potential due to the mass distribution.

A.6. A simple pendulum of mass m and length b oscillates in a plane about its equilibrium position. If θ is the angular displacement of the pendulum from equilibrium ($\theta = 0$), (a) write the Lagrangian of the system in terms of θ and b, and (b) use the Lagrange's equation to show that

$$\ddot{\theta} + \frac{g}{b} \sin\theta = 0.$$

Reference

Chow TL (1995) *Classical Mechanics*. (John Wiley & Sons, New York)

Appendix B
The Special Theory of Relativity

B.1 The Origins of Special Relativity

Before Einstein, the concept of space and time were those described by Galileo and Newton. Time was assumed to have an absolute or universal nature, in the sense that any two inertial observers who have synchronized their clocks will always agree on the time of any event (any happening that can be given a space and time coordinates). The Galilean transformations assert that any one inertial frame is as good as another one describing the laws of classical mechanics.

However, physicists of the 19th century were not able to grant the same freedom to electromagnetic theory, which did not seem to obey Galilean transformations. For example, the electromagnetism of Maxwell predicts that the speed of light is a constant, independent of the motion of the source and the observer. Now, a source at rest in an inertial frame S emits a light wave, which travels out as a spherical wave at a constant speed (3×10^5 km/sec). But observed in a frame S' moving uniformly with respect to S will see, according to Galilean transformations, that the light wave is no longer spherical and the speed of light is also different, so Maxwell's equations are not invariant under Galilean transformations. Therefore, for electromagnetic phenomena, inertial frames of reference are not equivalent.

Does this suggest that Maxwell's equations are wrong and need to be modified to obey the principle of Newtonian relativity? Or does this suggest that the existences of a preferred frame of reference in which Maxwell's equations are valid? The idea of a preferred frame of reference is foreign to classical mechanics. So a number of theories were proposed to resolve this conflict.

Today we know that Maxwell's equations are correct and have the same form in all inertial reference frames. There is some transformation other than the Galilean transformation that makes both electromagnetic and mechanical equations transform in an invariant way. But this proposal was not accepted without resistance. Owing to the works of Young and Fresnel, light was viewed as a mechanical wave, analogous to transverse waves on a string. Thus, its propagation required a physical medium. This medium was called ether and was required to have very strong restoring forces so that it could propagate light at such a great speed. But at the same

time the medium offers little resistance to the planets, as they suffered no observable reduction in speed even though they traveled through it year after year. It was
necessary to demonstrate the existence of the ether so that this paradox might be
resolved.

Since light can travel through space, it was assumed that ether must fill all of
space and the speed of light must be measured with respect to the stationary ether.

B.2 The Michelson-Morley Experiment

If ether does exist, it should be possible to detect some variation of the speed of light
as emitted by some terrestrial source. As Earth travels through space at 30 km/s in
an almost circular orbit around the sun, it is bound to have some relative velocity
with respect to ether. If this relative velocity is added to that of the light emitted
from the source, then light emitted simultaneously in two perpendicular directions
should be traveling at different speeds, corresponding to the two relative velocities
of the light with respect to the ether.

In 1887 Michelson set out to detect this velocity variation in the propagation of
light. His ingenious way of doing this depends on the phenomenon of interference
of light to determine whether the time taken for light to pass over two equal paths at
right angles was different or not. Figure B.1 shows schematically the interferometer
that Michelson used, which is essentially comprised of a light source S, a half-
silvered glass plate A, and two mirrors B and C, all mounted on a rigid base. The

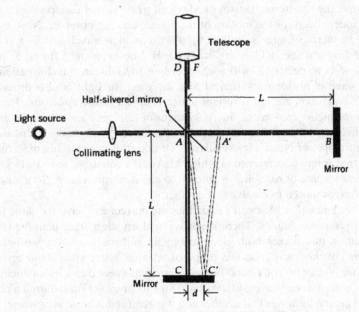

Fig. B.1 Schematic diagram of the Michelson-Morley Experiment.

two mirrors are placed at equal distances L from plate A. Light from S enters A and splits into two beams. One goes to mirror B, which reflects it back; the other beam goes to mirror C, also to be reflected back. On arriving back at A, the two reflected beams are recombined as two superimposed beams, D and F, as indicated. If the time taken for light to travel from A to B and back equals the time from A to C and back, the two beams D and F will be in phase and will reinforce each other. But if the two times differ slightly, the two beams will be slightly out of phase and interference will result. We now calculate the two times to see whether they are the same or not. We first calculate the time required for the light to go from A to B and back. If the line AB is parallel to Earth's motion in its orbit, and if Earth is moving at speed u and the speed of light in the ether is c, the time is

$$t_1 = \frac{L_{AB}}{c-u} + \frac{L_{AB}}{c+u} = \frac{2L_{AB}}{c[1-(u/c)^2]} \approx \frac{2L_{AB}}{c}\left(1 + \frac{u^2}{c^2}\right) \tag{A2-1}$$

where $(c-u)$ is the upstream speed of light with respect to the apparatus, and $(c+u)$ is the downstream speed.

Our next calculation is of the time t_2 for the light to go from A to C. We note that while light goes from A to C, the mirror C moves to the right relative to the ether through a distance $d = ut_2$ to the position C'; at the same time the light travels a distance ct_2 along AC'. For this right triangle we have

$$(ct_2)^2 = L_{AC}^2 + (ut_2)^2$$

from which we obtain

$$t_2 = \frac{L_{AC}}{\sqrt{c^2 - u^2}}.$$

Similarly, while the light is returning to the half-silvered plate, the plate moves to the right to the position B'. The total path length for the return trip is the same, as can be seen from the symmetry of Figure B.1. Therefore if the return time is also the same, the total time for light to go from A to C and back is then $2t_2$, which we denote by t_3:

$$t_3 = \frac{2L_{AC}}{\sqrt{c^2 - u^2}} = \frac{2L_{AC}}{c\sqrt{1-(u/c)^2}} \approx \frac{2L_{AC}}{c}\left(1 + \frac{u^2}{2c^2}\right). \tag{A2-2}$$

In (A2-1) and (A2-2) the first factors are the same and represent the time that would be taken if the apparatus were at rest relative to the ether. The second factors represent the modifications in the times caused by the motion of the apparatus. Now the time difference Δt is

$$\Delta t = t_3 - t_1 = \frac{2(L_{AC} - L_{AB})}{c} + \frac{L_{AC}}{c}\beta^2 - \frac{2L_{AB}}{c}\beta^2 \tag{A2-3}$$

where $\beta = u/c$.

It is most likely that we cannot make $L_{AB} = L_{AC} = L$ exactly. In that case we can rotate the apparatus 90 degrees, so that AC is in the line of motion and AB is perpendicular to the motion. Small differences in length become unimportant. Now we have

$$\Delta t' = t'_3 - t'_1 = \frac{2(L_{AB} - L_{AC})}{c} + \frac{2L_{AB}}{c}\beta^2 - \frac{L_{AC}}{c}\beta^2. \qquad \text{(A2-4)}$$

Thus,

$$\Delta t' - \Delta t = \frac{(L_{AB} + L_{AC})}{c}\beta^2. \qquad \text{(A2-5)}$$

This difference yields a shift in the interference pattern across the crosshairs of the viewing telescope. If the optical path difference between the beams changes by one λ (wave-length), for example, there will be a shift of one fringe. If δ represents the number of fringes moving past the crosshairs as the pattern shifts, then

$$\delta = \frac{c(\Delta t' - \Delta t)}{\lambda} = \frac{L_{AB} + L_{AC}}{\lambda}\beta^2 = \frac{\beta^2}{\lambda/(L_{AB} + L_{AC})}. \qquad \text{(A2-6)}$$

In the Michelson-Morley experiment of 1887, the effective length L was 11 m; sodium light of $\lambda = 5.9 \times 10^{-5}$ cm was used. The orbit speed of Earth is 3×10^4 m/s, so $\beta = 10^{-4}$. From (A2-6) the expected shift would be about 4/10 of a fringe

$$\delta = \frac{22m \times (10^{-4})^2}{5.9 \times 10^{-5}} = 0.37. \qquad \text{(A2-7)}$$

The Michelson-Morley interferometer can detect a shift of 0.005 fringes. However, no fringe shift in the interference pattern was observed. So no effect at all due to Earth's motion through the ether was found. This null result was very puzzling and most disturbing at the time. It was suggested, including by Michelson, that the ether might be dragged along by Earth, eliminating or reducing the ether wind in the laboratory. This is hard to square with the picture of the ether as an all-pervasive, frictionless medium. The ether's status as an absolute reference frame was also gone forever. Many attempts to save the ether failed (see Resnick and Halliday 1985). We just mention one here, namely the contraction hypothesis.

George F. Fitzgerald pointed out in 1892 that a contraction of bodies along the direction of their motion through the ether by a factor $(1 - u^2/c^2)^{1/2}$ would give the null result. Because (A2-1) must be multiplied by the contraction factor $(1 - u^2/c^2)^{1/2}$, then (A2-2) reduces to zero. The magnitude of this time difference is completely unaffected by rotation of the apparatus through 90°.

Lorentz obtained a contraction of this sort in his theory of electrons. He found that the field equations of electron theory remain unchanged if a contraction by the factor $(1 - u^2/c^2)^{1/2}$ takes places, provided also that a new measure of time is used in a uniformly moving system. The outcome of the Lorentz theory is that an observer will observe the same phenomena, no matter whether the person is at rest in the ether or moving with velocity. Thus, different observers are equally unable to tell whether they are at rest or moving in the ether. This means that for optical phenomena, just as for mechanics, ether is unobservable.

Poincaré offered another line of approach to the problem. He suggested that the result of the Michelson-Morley experiment was a manifestation of a general principle that absolute motion cannot be detected by laboratory experiments of any kind, and the laws of nature must be the same in all inertial reference frames.

B.3 The Postulates of the Special Theory of Relativity

Einstein realized the full implications of the Michelson-Morley experiment, the Lorentz theory, and Poincaré's principle of relativity. Instead of trying to patch up the accumulating difficulties and contradictions connected with the notion of ether, Einstein rejected the ether idea as unnecessary or unsuitable for the description of the physical laws. Along with the exit of ether, gone also was the notion of absolute motion through space. The Michelson-Morley experiment proved unequivocally that no such special frame of reference exists. All frames of reference in uniform relative motion are equivalent, for mechanical motions and also for electromagnetic phenomena. Einstein further extended this as a fundamental postulate, now known as the principle of relativity. Furthermore, he argued that the speed of light, c, predicted by electromagnetic theory must be a universal constant, the same for all observers. He took an epoch-making step in 1905 and developed the Special Theory of Relativity from these two basic postulates (assumptions), which are rephrased as follows:

1. *The laws of physics are the same in all inertial frames. No preferred inertial frame exists (the principle of relativity).*
2. *The velocity of light in free space is the same in all inertial frames and is independent of the motion of the emitting body (the principle of the constancy of the velocity of light).*

According to Einstein, sometime in 1896, after he entered the Zurich Polytechnic Institute to begin his education as a physicist, he asked himself the question of what would happen if he could catch up to a light ray—that is, move at the speed of light. Maxwell's theory says that light is a wave of electric and magnetic fields moving through space. But if you could catch up to a light wave, then the light wave would not be moving relative to you but instead be standing still. The light wave would then be a standing wave of electric and magnetic fields, which is not allowed if Maxwell's theory is right. So, he reasoned, there must be something wrong with the assumption that you can catch a light wave the same way as you can catch a water wave. This idea was the seed from which the fundamental postulate of the constancy of the speed of light and the Special Theory of Relativity grew nine years later.

All the seemingly very strange results of special relativity came from the special nature of the speed of light. Once we understand this, everything else in relativity makes sense. So let us take a brief look at the special nature of the speed of light. The speed of light is very great, 186,000 mi/s or 3×10^5 km/sec. But the bizarre fact of the speed of light is that it is independent of the motion of the observer or the source emitting the light. Michelson hoped to determine the absolute speed of Earth through ether by measuring the time differences required for light to travel across equal distances that are at right angles to each other. What did he observe? No difference in travel times for the two perpendicular light beams. It was as if Earth were absolutely stationary. The conclusion is that the speed of light does not depend on the motion of the object. This bizarre nature is not something that would be expected from common sense. The same common sense once told us that it was nonsense to think that Earth was round. So common sense is not always right!

How do we know that the speed of light is independent of the motion of the light source? There are many binary star systems in our galaxy, in which two stars revolve around a common center of mass. If the speed of light depended on the motion of the source, then the light emitted by the two stars in a binary system would have different speeds as they moved toward Earth, as shown in Figure B.2. The orbit is roughly edge-on to our line of sight, and its orbital speed about the center of mass of the system. If the distance to the binary system were right, we would receive light from the star at position A at the same time as the light sent to us at a slower speed and at an earlier time, when the star was at position B. Thus, under some circumstances we could be seeing the same star in a binary system at many different places in its orbit at once, and there would be multiple images or spread out images. But, in fact, we always see binary stars moving in a well-behaved elliptical orbit about each other. Thus, the motion of its source (the emitter) does not affect the speed of light.

Einstein's two postulates radically revised our concepts of space and time. Newtonian mechanics abolished the notion of absolute space. Now absolute space is abolished in its Maxwellian role as the ether, the carrier of electromagnetic waves. Time is also not absolute any more either, since all inertial observers agree on how fast light travels but not on how far light travels. Time has lost its universal nature. In fact, we shall see later examples of moving clocks that run slow. This is known as *time dilation*.

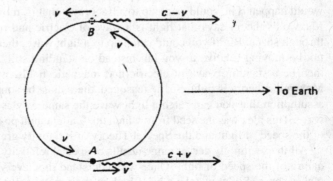

Fig. B.2 The nonexistence of light intensity variation from a binary star proves that the speed of light is independent of the motion of the light source.

B.4 The Lorentz Transformations

Since the Galilean transformations are inconsistent with Einstein's postulate of the constancy of the speed of light, we must modify it in such a way that the new transformation will incorporate Einstein's two postulates and make both mechanical and electromagnetic equations transforming in an invariant way. To this end we consider two inertial frames S and S'. Let the corresponding axes of S and S' frames be parallel, with frame S' moving at a constant velocity u relative to S along the x_1-axis. The apparatus for measurement of distances and times in the two frames are assumed identical, and the clocks are adjusted to read zero at the moment the two origins coincide. Figure B.3 represents the viewpoint of observers in S.

Suppose that an event occurred in frame S at the coordinates (x, y, z, t) and is observed at (x', y', z', t') in frame S'. Because of the homogeneity of space and time, we expect the transformation relations between the coordinates (x, y, z, t) and (x', y', z', t') to be linear, for otherwise there would not be a simple one-to-one relation between events in S and S' frames. For instance, a nonlinear transformation would predict acceleration in one system even if the velocity were constant in the other, obviously an unacceptable property for a transformation between inertial systems.

Let us consider the transverse dimensions first. Since the relative motion of the coordinate systems occurs only along the x-axis, we expect the linear relations are of the forms $y' = k_1 y$, $z' = k_2 z$. The symmetry requires that $y = k_1 y'$ and $z = k_2 z'$. These can both be true only if $k_1 = 1$ and $k_2 = 1$. Therefore, for the transverse direction we have

$$y' = y, \ z' = z. \tag{A2-8}$$

These relations are the same as in Galilean transformations.

Along the longitudinal dimension, the relation between x and x' must depend on the time, so let's consider the most general linear relation

$$x' = ax + bt. \tag{A2-9}$$

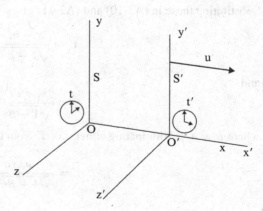

Fig. B.3 Relative motion of two inertial frames of reference.

Now, the origin O$'$, where $x' = 0$, corresponds to $x = ut$. Substituting these into (A2-9), we have

$$0 = aut + bt$$

from which we obtain

$$b = -au$$

and (A2-9) simplifies to

$$x' = a(x - ut). \tag{A2-10}$$

By symmetry, we also have

$$x = a(x' + ut') \tag{A2-11}$$

Now we apply Einstein's second postulate of the constancy of the speed of light. If a pulse of light is sent out from the origin O of frame S at $t = 0$, its position along the x-axis later is given by $x = ct$, and its position along x'-axis is $x' = ct'$. Putting these in (A2-10) and (A2-11), we obtain

$$ct' = a(c - u)t \quad \text{and} \quad ct = a(c + u)t'.$$

From these we obtain

$$\frac{t}{t'} = \frac{c}{a(c - u)} \quad \text{and} \quad \frac{t}{t'} = \frac{a(c + u)}{c}.$$

Therefore,

$$\frac{c}{a(c - u)} = \frac{a(c + u)}{c}.$$

Solving for a

$$a = \frac{1}{\sqrt{1 - (u/c)^2}},$$

then

$$b = -au = -\frac{u}{\sqrt{1 - (u/c)^2}}.$$

Substituting these in (A2-10) and (A2-11) gives

$$x' = \frac{x - ut}{\sqrt{1 - \beta^2}} \tag{A2-12a}$$

and

$$x = \frac{x' + ut'}{\sqrt{1 - \beta^2}} \tag{A2-12b}$$

where $\beta = u/c$. Eliminating either x or x' from (A2-12a) and (A2-12b), we obtain

$$t' = \frac{t - ux/c^2}{\sqrt{1 - \beta^2}} \tag{A2-12c}$$

and

$$t = \frac{t' + ux'/c^2}{\sqrt{1 - \beta^2}} \qquad \text{(A2-12d)}$$

Combining all of these results, we obtain the Lorentz transformations

$$\left.\begin{array}{ll} x' = \gamma\,(x - ut) & x = \gamma\,(x' + ut) \\ y' = y & y = y' \\ z' = z & z = z' \\ t' = \gamma\,(t - ux/c^2) & t = \gamma\,(t' + ux/c^2) \end{array}\right\} \qquad \text{(A2-13)}$$

where

$$\gamma\,(= 1/\sqrt{1 - \beta^2}),\ \ \beta = u/c \qquad \text{(A2-14)}$$

is the Lorentz factor.

If $\beta \ll 1$, then $\gamma \cong 1$, and (A2-13) reduces to the Galilean transformations. That is, the Galilean transformations are a first approximation to the Lorentz transformations for $\beta \ll 1$.

When the velocity, $\vec{u}$, of S' relative to S is in some arbitrary direction, (A2-13) can be given a more general form in terms of the components of $\vec{r}$ and $\vec{r}'$ perpendicular and parallel to $\vec{u}$.

$$\left.\begin{array}{ll} \vec{r}_{\parallel}' = \gamma\,(\vec{r}_{\parallel} - \vec{u}t) & \vec{r}_{\parallel} = \gamma\,(\vec{r}_{\parallel}' + \vec{u}t) \\ \vec{r}_{\perp}' = \vec{r}_{\perp} & \vec{r}_{\perp}' = \vec{r}_{\perp} \\ t' = \gamma\,(t - \vec{u} \cdot \vec{r}/c^2) & t = \gamma\,(t' + \vec{u} \cdot \vec{r}/c^2) \end{array}\right\} \qquad \text{(A2-15)}$$

The Lorentz transformations are valid for all types of physical phenomena at all speeds. As a consequence of this all physical laws must be invariant under a Lorentz transformation.

The Lorentz transformations that are based on Einstein's postulates contain a new philosophy of space and time measurements. We now examine the various properties of these new transformations. In the following discussion, we still use Figure B.3.

B.4.1 Relativity of Simultaneity and Causality

Two events that happen at the same time but not necessarily at the same place are called simultaneous. Now consider two events in S' that occur at (x_1', t_1') and (x_2', t_2'); they would appear in frame S at (x_1, t_1) and (x_2, t_2). The Lorentz transformations give

$$t_2 - t_1 = \gamma \left[(t_2' - t_1') + \frac{u(x_2' - x_1')}{c^2} \right]. \qquad \text{(A2-16)}$$

Now, it is easy to observe that if the two events take place simultaneously in S' (so $t_2' - t_1' = 0$), they do not occur simultaneously in the S frame, for there is a finite time lapse

$$\Delta t = t_2 - t_1 = \gamma \, \frac{u(x_2' - x_1')}{c^2} \neq 0.$$

Thus, two spatially separated events that are simultaneous in S' would not be simultaneous in S. In other words, the simultaneity of spatially separated events is not an absolute property, as it was assumed to be in Newtonian mechanics. Moreover, depending on the sign of $(x_2' - x_1')$ the time interval Δt can be positive or negative, that is, in the frame S the "first" event in S' can take place earlier or later than the "second" one. The exception is the case when two events occur coincidentally in S'; then they also occur at the same place and at the same time in frame S.

If the order of events in frame S is not reversed in time, then $\Delta t = t_2 - t_1 > 0$, which implies that

$$(t_2' - t_1') + \frac{u(x_2' - x_1')}{c^2} > 0$$

or

$$\frac{x_2' - x_1'}{t_2' - t_1'} < \frac{c^2}{u}$$

which will be true as long as

$$\frac{x_2' - x_1'}{t_2' - t_1'} < c.$$

Thus the order of events will remain unchanged if no signal can be transmitted with a speed greater than c, the speed of light.

B.4.2 Time Dilation and Relativity of Co-locality

Two events that happen at the same place but not necessarily at the same time are called *co-local*. Now consider two co-local events in S' taking place at t_1' and t_2' but at the same place. For simplicity consider this to be on the x'-axis so that $y' = z' = 0$. These two events would appear in frame S at (x_1, t_1) and (x_2, t_2). The Lorentz transformations give

$$\Delta x = \frac{u \Delta t'}{\sqrt{1 - \beta^2}} = \gamma u \Delta t', \quad \Delta t = \frac{\Delta t'}{\sqrt{1 - \beta^2}} = \gamma \Delta t' \qquad \text{(A2-17)}$$

where $\beta = u/c$, $\Delta t' = t_2' - t_1'$, and so forth. It is easy to observe that:

(1) Two co-local events in S' do not occur at the same place in S, and so t_1 and t_2 must be measured by spatially separated synchronized clocks. Einstein's prescription for synchronizing two stationary separated clocks is to send a light signal from clock 1 at a time t_1 (measured by clock 1) and reflected back from clock 2 at a time t_2 (measured by clock 2). If the reflected light returns to clock 1 at a time t_3 (measured by clock 1), then clocks 1 and 2 are synchronous if $t_2 - t_1 = t_3 - t_2$; that is, if the time measured for light to go one way is equal to the time measured for light to go in the opposite direction.

(2) The time interval between two co-local events in an inertial reference is measured by a single clock at a given point and it is called the *proper time* interval between the two events. In the second equation of (A2-17), $\Delta t' = t_2' - t_1'$ is the proper time interval between the events in S'. Since $\gamma \geq 1$, the time interval $\Delta t = t_2 - t_1$ in S is longer than $\Delta t'$; this is called time dilation, often described by the statement that "moving clocks run slow." This apparent asymmetry between S and S' in time is a result of the asymmetric nature of the time measurement.

Time dilation has been confirmed by experiments on the decay of pions. Pions have a mean lifetime of $T_0 = 2.6 \times 10^{-8}$ sec when they are at rest. When they are in fast motion in a synchrotron, their lifetimes become larger according to

$$T = \frac{T_0}{\sqrt{1 - \beta^2}}.$$

Time dilation between observers in uniform relative motion is a very real thing. All processes, including atomic and biological processes, slow down in moving systems.

We often hear the twin paradox. Consider one twin gets on a spaceship and accelerates to $0.866c$, so $\gamma = 2$. If this twin travels for one year, as measured by his clock, then heads back at the same speed, the moving twin will report that the trip required two years. But his Earth-bound twin would report that the time for the journey was four years. Can the twin on the spaceship argue that he was at rest, and it was the twin on earth who was moving? The answer is no. To make a transition from a rest frame to a moving frame and to turn around heading for home, there must be acceleration. The twin who feels acceleration can no longer claim that his frame is the rest frame. Thus the twin on the spaceship cannot argue that it was his Earth-bound twin moving, and there is no paradox.

B.4.3 Length contraction

Consider a rod of length L_0 lying at rest along the x' axis in the S' frame: $L_0 = x_2' - x_1'$. L_0 is the proper length of the rod measured in the rod's rest frame S'. Now the rod is moving lengthwise with velocity u relative to the S frame. An observer in the S frame makes a simultaneous measurement of the two ends of the rod. The Lorentz transformations give

$$x_1' = \gamma [x_1 - ut_1], \quad x_2' = \gamma [x_2 - ut_2]$$

from which we get

$$x_2' - x_1' = \gamma [x_2 - x_1] - \gamma u(t_2 - t_1) = \gamma [x_2 - x_1],$$

where we dropped the $(t_2 - t_1)$ term, because the measurement in S is made simultaneously. The above result often is rewritten as

$$L(u) = \sqrt{1 - \beta^2} L_0 \qquad \qquad \text{(A2-18)}$$

where $L(u) = x_2 - x_1$. Thus the length of a body moving with velocity u relative to an observer is measured to be shorter by a factor of $(1 - \beta^2)^{1/2}$ in the direction of motion relative to the observer.

Since all inertial frames are equally valid, if $L_0 = \gamma L$, is the expression $L = \gamma L_0$ to be equally true? The answer is no, because the measurement was not carried out in the same way in the two reference frames. The positions of the two ends of the rod were marked simultaneously in the S frame, but they are not simultaneous in the S' frame. This difference gives the asymmetry of the result. As a general expression, $\Delta x' = \gamma \Delta x$ is not true. The full expression relating distances in two frames of reference is $\Delta x' = \gamma(\Delta x - u \Delta t)$, and the symmetrical inverse relation is $\Delta x = \gamma(\Delta x' + u \Delta t')$. In the case that was considered earlier, $\Delta t = 0$, so $\Delta x' = \gamma \Delta x$, but $\Delta t' \neq 0$, so $\Delta x \neq \gamma \Delta x'$.

A body of proper volume V_0 can be divided into thin rods parallel to u. Each one of these rods is reduced in length by a factor $(1 - \beta^2)^{1/2}$ so that the volume of the moving body measured by an observer in S is $V = (1 - \beta^2)^{1/2} V_0$.

An interesting consequence of the length contraction is the visual apparent shape of a rapidly moving object. This was shown first by James Terrell in 1959 [*Physics Review*, 116(1959), 1041; and *American Journal of Physics*, 28 (1960) 607]. The act of seeing involves the simultaneous light reception from different parts of the object. In order for light from different parts of an object to reach the eye or a camera at the same time, light from different parts of the object must be emitted at different times, to compensate for the different distances the light must travel. Thus, taking a picture of a moving object or looking at it does not give a valid impression of its shape. Interestingly, the distortion that makes the Lorentz contraction seem to disappear instead makes an object seem to rotate by an angle $\theta = \sin^{-1}(u/c)$, as long as the angle subtended by the object at the camera is small. If the object moves in another direction, or if the angle it subtends at the camera is not small, the apparent distortion becomes quite complex.

Figure B.4 shows a cube of side l moving with a uniform velocity u with respect to an observer some distance away; the side AB is perpendicular to the line of sight of the observer. In order for light from corners A and D to reach the observer at the same instant, light from D, which must travel a distance l farther than from A, must have been emitted when D was in position E. The length DE is equal to $(l/c)u = l\beta$. The length of the side AB is foreshortened by Lorentz contraction to $l\sqrt{1 - \beta^2}$. The net result corresponds to the view the observer would have if the cube were rotated through an angle $\sin^{-1}\beta$. The cube is not distorted; it undergoes an apparent rotation. Similarly, a moving sphere will not become an ellipsoid; it still appears as a sphere. Weisskopf (*Physics Today* 13, **9**, 1960) gives an interesting discussion of apparent rotations at high velocity.

Length contraction opens the possibility of space travel. The nearest star, besides the sun, is Alpha Centauri, that is about 4.3 light-years away; light from Alpha Centauri takes 4.3 years to reach us. Even if a spaceship can travel at the speed of light, it would take 4.3 years to reach Alpha Centauri. This is certainly true from the

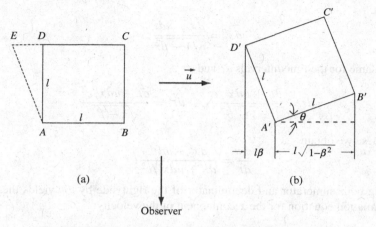

Observer

Fig. B.4 A rapidly moving object undergoes an apparent rotation.

point of view of an observer on Earth. But from the point of view of the crew of the spaceship, the distance between Earth and Alpha Centauri is shortened by a factor $\gamma = (1 - \beta^2)^{1/2}$, where $\beta = v/c$ and v is the speed of the spaceship. If v is, say, $0.99c$, then $\gamma = 0.14$, and the distance appears to be only 14% of the value as seen from Earth. If the crew, therefore, deduces that light from Alpha Centauri takes only $0.14 \times 4.3 = 0.6$ year to reach earth and sees Alpha Centauri coming toward them at a speed of $0.99c$, they expect to get there in $0.60/0.99 = 0.606$ years, without having to suffer a long tedious journey. But, in practice, the power requirements to launch a spaceship near the speed of light are prohibitive.

B.4.4 Velocity Transformation

The new and more complicated transformation for velocities can be deduced easily from Lorentz transformations. By definition the components of velocity in S and S′ frames are given by, respectively,

$$v_x = \frac{dx}{dt} = \frac{x_2 - x_1}{t_2 - t_1}, \quad v_x' = \frac{dx'}{dt} = \frac{x'_2 - x'_1}{t'_2 - t'_1},$$

$$v_y = \frac{dy}{dt} = \frac{y_2 - y_1}{t_2 - t_1}, \quad v_y' = \frac{dy'}{dt} = \frac{y'_2 - y'_1}{t'_2 - t'_1}, \text{ and so on.}$$

Applying the Lorentz transformations to x_1 and x_2 and then taking the difference, we get

$$dx = \frac{dx' + u\,dt'}{\sqrt{1 - \beta^2}}, \quad \beta = \frac{u}{c}.$$

Similarly,

$$dx' = \frac{dx - udt}{\sqrt{1 - \beta^2}}.$$

Do the same for the time intervals dt and dt':

$$dt = \frac{dt' + udx'/c^2}{\sqrt{1 - \beta^2}}, \quad dt' = \frac{dt - udx/c^2}{\sqrt{1 - \beta^2}}.$$

From these we obtain

$$\frac{dx}{dt} = \frac{dx' + udt'}{dt' + udx'/c^2}.$$

Dividing both numerator and denominator of the right side by dt' yields the right transformation equation for the x component of the velocity:

$$v_x = \frac{v'_x + u}{1 + uv'_x/c^2}. \tag{A2-19a}$$

Similarly, we can find the transverse components:

$$v_y = \frac{v'_y\sqrt{1 - \beta^2}}{1 + uv'_x/c^2} = \frac{v'_y}{\gamma(1 + uv'_x/c^2)} \tag{A2-19b}$$

$$v_z = \frac{v'_z\sqrt{1 - \beta^2}}{1 + uv'_x/c^2} = \frac{v'_z}{\gamma(1 + uv'_x/c^2)}. \tag{A2-19c}$$

In these formulas, $\gamma = (1 - \beta^2)^{-1/2}$ as before. We note that the transverse velocity components depend on the x-component. For $v \ll c$, we obtain the Galilean result $v_x = v_x' + u$. Solving explicitly or merely switching the sign of u would yield (v_x', v_y', v_z') in terms of (v_x, v_y, v_z).

It follows from the velocity transformation formulas that the value of an angle is relative and changes in transition from one reference frame to another. For an object in the S frame moving in the xy-plane with velocity v that makes an angle θ with the x-axis, we have

$$\tan\theta = v_y/v_x, \quad v_x = v\cos\theta, \quad v_y = v\sin\theta$$

In the S' frame, we have

$$\tan\theta' = \frac{v_y'}{v_x'} = \frac{v\sin\theta}{\gamma(v\cos\theta - u)} \tag{A2-20}$$

where $\gamma = 1/\sqrt{1 - \beta^2}$, and $\beta = u/c$.

As an application, consider the case of starlight, that is, $v = c$; then

$$\tan\theta' = \frac{\sin\theta}{\gamma(\cos\theta - u/c)}.$$

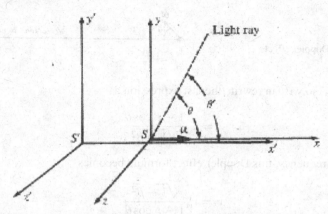

Fig. B.5 Aberration. The angles of a light ray with x-axis and x'-axis are different.

Let $\theta = \pi/2, \theta' = \pi/2 - \phi$ (Fig. B.5); from this we obtain the star aberration formula, to see a star overhead tilt the telescope at angle ϕ:

$$\tan\phi = \frac{-u/c}{\sqrt{1-\beta^2}}, \quad \sin\phi = -\frac{u}{c}$$

B.5 The Doppler Effect

The Doppler effect occurs for light as well as for sound. It is a shift in frequency due to the motion of the source or the observer. Knowledge of the motion of distant receding galaxies comes from studies of the Doppler shift of their spectral lines. The Doppler effect is also used for satellite tracking and radar speed traps. We examine the Doppler effect in light only.

Consider a source of light or radio waves moving with respect to an observer or a receiver, at a speed u and at an angle θ with respect to the line between the source and the observer (Fig. B.6). The light source flashes with a period τ_0 in its rest frame (the S' frame in which the source is at rest). The corresponding frequency is $\nu_0 = 1/\tau_0$, and the wavelength is $\lambda_0 = c/\nu_0 = c\tau_0$.

While the source is going through one oscillation, the time that elapses in the rest frame of the observer (the S frame) is $\tau = \gamma\tau_0$ because of time dilation, where $\gamma = (1-\beta^2)^{1/2}$ and $\beta = u/c$. The emitted wave travels at speed c, and therefore its front moves a distance of $\gamma\tau_0 c$; the source moves toward the observer with a speed $u\cos\theta$, so a distance of $\gamma\tau_0 u\cos\theta$. Then the distance D separates the fronts of the successive waves (the wavelength):

$$D = \gamma\tau_0 c - \gamma\tau_0 u\cos\theta,$$

i.e.,

$$\lambda = \gamma\tau_0 c - \gamma\tau_0 u\cos\theta = \gamma\tau_0 c[1 - (u/c)\cos\theta].$$

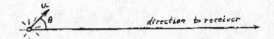

Fig. B.6 The Doppler effect.

but $c\tau_0 = \lambda_0$, so we can rewrite the last expression as

$$\lambda = \lambda_0 \frac{1 - \beta \cos\theta}{\sqrt{1 - \beta^2}}. \tag{A2-21}$$

In terms of frequency, this Doppler effect formula becomes

$$\nu = \nu_0 \frac{\sqrt{1 - \beta^2}}{1 - \beta \cos\theta}. \tag{A2-22}$$

Here ν is the frequency at the observer, and θ is the angle measured in the rest frame of the observer. If the source is moving directly toward the observer, then $\theta = 0$ and $\cos\theta = 1$. (A2-22) reduces to

$$\nu = \nu_0 \frac{\sqrt{1 - \beta^2}}{1 - \beta} = \nu_0 \sqrt{\frac{1 + \beta}{1 - \beta}}. \tag{A2-23a}$$

For a source moving directly away from the observer, $\cos\theta = -1$, (A2-24) reduces to

$$\nu = \nu_0 \frac{\sqrt{1 - \beta^2}}{1 + \beta} = \nu_0 \sqrt{\frac{1 - \beta}{1 + \beta}}. \tag{A2-23b}$$

At $\theta = \pi/2$, i.e., the source moving at right angles to the direction of the observer (A2-24) and reduces to

$$\nu = \nu_0 \sqrt{1 - \beta^2}. \tag{A2-23c}$$

This transverse Doppler effect is due to time dilation.

B.6 Relativistic Space-Time and Minkowski Space

In our daily experience we are used to thinking of a world of three dimensions. Objects in space have length, breadth, and height. We tend to think of time as being independent of space. However, as we have just seen, there is no absolute standard for the measurement of time or of space; the relative motion of observers affects both kinds of measurement. Lorentz transformations treat x^i ($i = 1, 2, 3$) and t as equivalent variables. In 1907 H. Minkowski proposed that the three dimensions of space and the dimension of time should be treated together as a fourth dimension of space-time. Minkowski remarked: "Henceforth space by itself and time by itself are doomed to fade away into mere shadows, and only a kind of union of the two will preserve an independent reality." And he called the four dimensions of space-time the world space, and the path of an individual particle in space-time a world line. The four-dimension relativistic space-time is often called the Minkowski space.

It is now a common practice to treat t as a zeroth or a fourth coordinate

$$x^0 = ct, x^1 = x, x^2 = y, x^3 = z \tag{A2-24a}$$

or

$$x^1 = x, x^2 = y, x^3 = z, x^4 = ix^0. \tag{A2-24b}$$

By analogy with the three-dimensional case, the coordinates of an event (x^0, x^1, x^2, x^3) can be considered as the components of a four-dimensional radius vector, for short, a radius four-vector in Minkowski space. The square of the length of the radius four-vector is given by

$$(x^1)^2 + (x^2)^2 + (x^3)^2 + (x^4)^2 = -[(x^0)^2 - (x^1)^2 - (x^2)^2 - (x^3)^2].$$

It does not change under Lorentz transformations.

The Lorentz transformations now take on the form

$$
\begin{aligned}
x'^1 &= \gamma (x^1 + i\beta x^4) & x'^0 &= \gamma (x^0 - \beta x^1) \\
x'^2 &= x^2 & x'^1 &= \gamma (-\beta x^0 + x^1) \\
x'^3 &= x^3 \quad \text{or} & x'^2 &= x^2 \\
x'^4 &= \gamma (-i\beta x^1 + x^4) & x'^3 &= x^3.
\end{aligned} \tag{A2-25}
$$

In matrix form, we have

$$
\begin{pmatrix} x'^1 \\ x'^2 \\ x'^3 \\ x'^4 \end{pmatrix} = \begin{pmatrix} \gamma & 0 & 0 & i\beta\gamma \\ 0 & 1 & 0 & 0 \\ 0 & 0 & 1 & 0 \\ -i\beta\gamma & 0 & 0 & \gamma \end{pmatrix} \begin{pmatrix} x^1 \\ x^2 \\ x^3 \\ x^4 \end{pmatrix} \quad \text{or} \quad \begin{pmatrix} x'^0 \\ x'^1 \\ x'^2 \\ x'^3 \end{pmatrix} = \begin{pmatrix} \gamma & -\beta\gamma & 0 & 0 \\ -\beta\gamma & \gamma & 0 & 0 \\ 0 & 0 & 1 & 0 \\ 0 & 0 & 0 & 1 \end{pmatrix}. \tag{A2-26}
$$

We will use Greek indices (μ and ν, etc.) to label four-dimensional variables and Latin indices (i and j, etc.) to label three-dimensional variables.

The Lorentz transformations can be distilled into a single equation

$$x'^\mu = \sum_{\nu=1}^{4} L_\nu^\mu x^\nu = L_\nu^\mu x^\nu \quad \mu, \nu = 1, 2, 3, 4 \tag{A2-27}$$

where L_ν^μ is the Lorentz transformation matrix in (A2-26). The summation sign is eliminated in the last step by Einstein summation convention; the repeated indexes appearing once in the lower and once in the upper position are summed over. However, the indexes repeated in the lower part or upper part alone are not summed over.

If (A2-27) reminds you of the orthogonal rotations, it is no accident! The general Lorentz transformations can indeed be interpreted as an orthogonal rotation of axes in Minkowski space. The xt-submatrix of the Lorentz matrix in (A2-26) is

$$\begin{pmatrix} \gamma & i\beta\,\gamma \\ -i\beta\,\gamma & \gamma \end{pmatrix};$$

let us compare it with the xy-submatrix of the two-dimensional rotation about the z-axis

$$\begin{pmatrix} \cos\theta & \sin\theta \\ -\sin\theta & \cos\theta \end{pmatrix}.$$

Upon identification of matrix elements $\cos\theta = \gamma$, $\sin\theta = i\beta\gamma$, we see that the rotation angle θ (for the rotation in the xt-plane) is purely imaginary.

Some books prefer to use a real angle of rotation ϕ, defining $\phi = -i\theta$. Then note that

$$\cos\theta = \frac{e^{i\theta} + e^{-i\theta}}{2} = \frac{e^{-\phi} + e^{\phi}}{2} = \cosh\phi$$

$$\sin\theta = \frac{e^{i\theta} - e^{-i\theta}}{2i} = \frac{i[e^{\phi} - e^{-\phi}]}{2} = \sinh\phi$$

and the submatrix becomes

$$\begin{pmatrix} \cosh\phi & i\sinh\phi \\ -i\sinh\phi & \cosh\phi \end{pmatrix}.$$

We should note that the mathematical form of Minkowski space looks exactly like a Euclidean space; however, it is not physically so because of its complex nature as compared to the real nature of the Euclidean space.

B.6.1 Interval ds^2 as an Invariant

We are always interested in an invariant quantity that is unaffected by different choices of coordinate systems. We will see that intervals are Lorentz invariants. Let us consider again the two frames S and S′ in Figure B.3, moving relative to each other with constant velocity. The wave front of light that is emitted at the origin of frame S when $t = 0$ is given by

$$c^2t^2 - x^2 - y^2 - z^2 = (x^0)^2 - (x^1)^2 - (x^2)^2 - (x^3)^2 = 0. \qquad \text{(A2-28a)}$$

The wave front of the light will give a cone around the t axis. This is called the light cone (Figure B.7). The same wave front will have different coordinates

$$c^2t'^2 - x'^2 - y'^2 - z'^2 = (x'^0)^2 - (x'^1)^2 - (x'^2)^2 - (x'^3)^2 = 0. \qquad \text{(A2-28b)}$$

For any two events, such as sending out and receiving a light signal, the quantity s_{12}, where

$$s_{12} = \left[(x_2^0 - x_1^0)^2 - (x_2^1 - x_1^1)^2 - (x_2^2 - x_1^2)^2 - (x_2^3 - x_1^3)^2 \right]^{1/2} \qquad \text{(A2-29)}$$

is called the interval between the two events. (A2-28a) and (A2-28b) indicate that if the interval between two events is zero in one coordinate frame, it is also zero in all other frames.

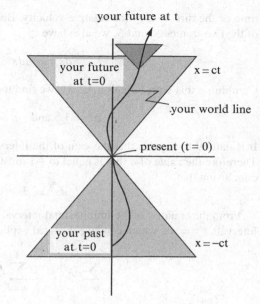

Fig. B.7 Spacetime diagram of a two-dimensional world, showing the light cone.

The interval ds between two events that are infinitesimally close to each other is

$$ds^2 = c^2 dt^2 - (dx^2 + dy^2 + dz^2)$$
$$= -\left[(dx^0)^2 + (dx^1)^2 + (dx^2)^2 + (dx^3)^2\right] \qquad \text{(A2-30)}$$

From the formal point of view, ds^2 can be regarded as the square of the distance between two world points in Minkowski space. We may rewrite ds^2 in a more general form

$$ds^2 = \sum_{\mu,\nu=0}^{3} g_{\mu\nu} dx^\mu dx^\nu \qquad \text{(A2-30a)}$$

where

$$g_{00} = 1, g_{11} = g_{22} = g_{33} = -1; \; g_{\mu\nu} = 0 \text{ if } \mu \neq \nu. \qquad \text{(A2-30b)}$$

The reader should be aware that the sign for the g's is not standard. Others may define $-s_{12}$ as the interval between two events; if so, then $g_{00} = -1 \; g_{11} = 1$ etc.

If $ds = 0$ in frame S, then $ds' = 0$ in frame S'. Furthermore, ds and ds' are infinitesimal of the same order. It follows that ds and ds' must be proportional to each other

$$ds^2 = ads'^2 \qquad \text{(A2-31)}$$

where the proportionality constant a may depend on the absolute value of the relative velocity of the two inertial frames. Owing to the homogeneity of space and time and the isotropy of space, the coefficient a cannot depend on the coordinates or the time or the direction of the relative velocity. Because of the complete equivalence of the two frames S' and S, we also have

$$ds'^2 = a\,ds^2. \tag{A2-32}$$

Combining this with Equation A2-31, we find that

$$a^2 = 1 \quad \text{and} \quad a = \pm 1.$$

It is natural to assume that the sign of the interval in all frames must be the same. Therefore the value of a that is equal to -1 must be discarded. We thus arrive at the conclusion that

$$ds^2 = ds'^2. \tag{A2-33}$$

From the equality of the infinitesimal intervals there follows the equality of finite intervals $s'^2 = s^2$, which can be expressed explicitly as

$$\sum_{\mu=0}^{3} (x'^\mu)^2 = \sum_{\mu=0}^{3} (x^\mu)^2. \tag{A2-34}$$

This invariance of the interval between two events is the mathematical expression of the constancy of the velocity of light.

Equation (A2-34) is analogous to three-dimensional length-preserving orthogonal rotations, and indicates again that Lorentz transformations corresponding to a rotation in Minkowski space.

The invariance of the interval ds^2 is a very useful tool in many of its applications. The skillful use of this invariance often avoids an explicit Lorentz transformation. Some insight into the nature of the interval is gained by considering some special cases. First, we introduce the notations

$$t_{12} = t_2 - t_1, \quad d_{12} = (x_2^1 - x_1^1)^2 - (x_2^2 - x_1^2)^2 - (x_2^3 - x_1^3)^2.$$

The interval between two events in frame S now takes a simpler appearance:

$$s_{12}{}^2 = c^2 t_{12}{}^2 - d_{12}{}^2.$$

If the two events occur at the same place in S$'$ frame, then $d'_{12} = 0$, and because of the invariance of the intervals, we have

$$s_{12}^2 = c^2 t_{12}^2 - d_{12}^2 = c^2 t'_{12}{}^2 > 0, \tag{A2-35}$$

and the interval is real. Real intervals are called timelike. The time interval between two events in S$'$ frame is

$$t'_{12} = \frac{s_{12}}{c} = \frac{\sqrt{c^2 t_{12}^2 - d_{12}^2}}{c}.$$

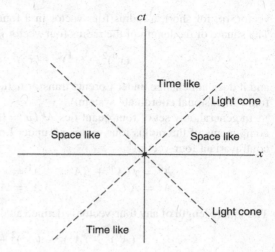

Fig. B.8 Spacetime intervals.

Every timelike interval that connects event 1 with another event lies within the light cones bounded by $x_1 = \pm ct$. All events that could have affected event 1 lie in the past light cone, and all events that event 1 is able to affect lie in the future light cone.

If the two events occur at one and the same time (simultaneously) in the S' frame, then $t'_{12} = 0$, and we have

$$s_{12}^2 = c^2 t_{12}^2 - d_{12}^2 = -d'_{12}{}^2 < 0, \qquad (A2\text{-}36)$$

and the interval is imaginary. Such an interval is called spacelike. There is no causal relationship between events 1 and 2. Every event that is connected with event 1 by a spacelike interval lies outside the light cone of event 1, and neither has interacted with event 1 in the past nor is capable of interacting with it in the future (Fig. B.8).

When two events can be connected with a light signal only, then

$$S_{12} = 0, \qquad (A2\text{-}37)$$

and such an interval is said to be lightlike. Events that can be connected with event 1 by lightlike intervals lie on the boundaries of the light cones.

The world line of a particle (the path of a particle in Minkowski space) must lie within its light cones. The division of intervals into spacelike, timelike, and lightlike intervals is, because of their invariance, an absolute concept. This means that the timelike, spacelike, or lightlike character of an interval is independent of the frame of reference.

B.6.2 Four Vectors

By analogy with the three-dimensional case, the coordinates of an event (x^0, x^1, x^2, x^3) can be considered as the components of a four-dimensional radius vector, or, for short, a radius four-vector in a four-dimensional Minkowski space. The square of the length of the radius four-vector is given by

$$(x^0)^2 - (x^1)^2 - (x^2)^2 - (x^3)^2$$

and it does not change under Lorentz transformations (or under any rotations of the four-dimensional coordinate system.)

In general, any set of four quantities $A^\mu (\mu = 0, 1, 2, 3)$ that transforms like the components of the radius four-vector x^μ under Lorentz transformations is called a contravariant four-vector:

$$\left.\begin{aligned}A^0 &= \gamma\,(A'^0 + \beta A'^1), \quad A^1 = \gamma\,(A'^1 + \beta A'^0)\\A^2 &= A'^2, \qquad\qquad\qquad\; A^3 = A'^3\end{aligned}\right\} \tag{A2-38}$$

The square length of any four-vector is defined analogously to the radius four-vector,

$$(A^0)^2 - (A^1)^2 - (A^2)^2 - (A^3)^2. \tag{A2-39}$$

The components of covariant four vectors A_μ are related to contravariant vectors by the following equation:

$$A_\mu = g_{\mu\nu} A^\nu \tag{A2-40}$$

where $g_{\mu\nu}$ is given by (A2-30b).

With the two types of four vectors, we can form the scalar product that is an invariant:

$$\sum_{\mu=0}^{3} A_\mu A^\mu = A_\mu A^\mu. \tag{A2-41}$$

The summation sign is eliminated in the last step by Einstein's summation convention. The $g_{\mu\nu}$ is a device to lower the indexes. Likewise, we can define $g^{\mu\nu}$ to raise indexes. In the Cartesian coordinates used here

$$g^{\mu\nu} = g_{\mu\nu}, \text{ and } g_{\mu\nu} g^{\mu\sigma} = \delta_\nu^\sigma \tag{A2-42}$$

where δ_μ^ν is the kronecker delta symbol, $\delta_\mu^\nu = 1$ if $\mu = \nu$, and $\delta_\mu^\nu = 0$ if $\mu \neq \nu$.

We can define quantities $A^{\mu\nu}$ or $A_{\mu\nu}$ which, for each index, behave like a vector. Evidently such a quantity transforms like

$$A'^{\mu\nu} = \frac{\partial x'^\mu}{\partial x^\alpha} \frac{\partial x'^\nu}{\partial x^\beta} A^{\alpha\beta} \tag{A2-43}$$

and is called a tensor of second rank or second-order tensor. A second-order tensor is said to be symmetric if $A^{\mu\nu} = A^{\nu\mu}$, and antisymmetric if $A^{\mu\nu} = -A^{\nu\mu}$. Tensors of higher rank are similarly defined:

$$A'^{\mu\nu\dots\sigma} = \frac{\partial x'^\mu}{\partial x^\alpha} \frac{\partial x'^\nu}{\partial x^\beta} \cdots \frac{\partial x'^\sigma}{\partial x^\alpha} A^{\alpha\beta\dots\lambda}.$$

A partial differential operator behaves like a vector. This can be seen from its transformation equation. From the chain rule of differential calculus we get

$$\frac{\partial}{\partial t} = \frac{\partial t'}{\partial t}\frac{\partial}{\partial t'} + \frac{\partial x'}{\partial t}\frac{\partial}{\partial x'}.$$

The coefficient can be read off from (A2-25),

$$\frac{\partial}{\partial t} = \gamma\frac{\partial}{\partial t'} + (-\beta\gamma c)\frac{\partial}{\partial x'} = \gamma\left(\frac{\partial}{\partial t'} - v\frac{\partial}{\partial x'}\right). \qquad \text{(A2-44a)}$$

Similarly

$$\frac{\partial}{\partial x} = \frac{\partial t'}{\partial x}\frac{\partial}{\partial t'} + \frac{\partial x'}{\partial x}\frac{\partial}{\partial x'} = \gamma\left(\frac{\partial}{\partial x'} - \frac{v}{c^2}\frac{\partial}{\partial t'}\right). \qquad \text{(A2-44b)}$$

For convenience, we will write $\partial_x = \partial/\partial x$, $\partial'_x = \partial/\partial x'$ and so on.

Since the differentiation $\partial_\mu = \partial/\partial x^\mu$ behaves as a vector, we can obtain new tensors by differentiating tensors, for example,

$$\text{gradient}: V_\mu = \partial_\mu\Phi, \qquad \text{curl}: f_{\mu\nu} = \partial_\mu A_\nu - \partial_\nu A_\mu,. \qquad \text{(A2-45a)}$$

$$\text{divergence}: \alpha = \partial_\mu A^\mu, \qquad \text{d'Alembertian}\ \lambda = \partial_\mu\partial^\mu\Phi. \qquad \text{(A2-45b)}$$

Under certain conditions, new tensors can also be formed by integration. To show this, consider the differentials

$$d\Omega = dx^0 dx^1 dx^2 dx^3 = cdtdV, \qquad\qquad dx^\mu = (dx^0, dx^1, dx^2, dx^3)$$
$$dS_\mu = (dx^2 dx^3 dx^0, dx^1 dx^3 dx^0, dx^1 dx^2 dx^0, dx^1 dx^2 dx^3), \qquad dS^{\mu\nu} = dx^\mu dx^\nu.$$

We can also construct the integral quantities, for example

$$\Lambda = \int\Phi d\Omega \quad \text{scalar}, \qquad\qquad \Lambda = \int A_\mu dx^\mu \quad \text{scalar}$$
$$\Lambda^\mu = \int T^{\mu\nu} dS_\nu, \quad \text{vector} \qquad \Lambda = \int U^\mu dS_\mu \quad \text{scalar, etc.}$$

Among these expressions, the following are important:

$$
\begin{array}{ll}
\oint A_\mu dx^\mu & \text{line integration} \\
\int dx^\mu dx^\nu B_{\mu\nu} & \text{two-surface integration} \\
\int A^\mu dS_\mu & \text{three-surface integration} \\
\int \Phi d\Omega & \text{space-time integration.}
\end{array}
$$

There are theorems that enable us to transform four-dimensional integrals, analogous to the theorems of Gauss and Stokes in three-dimensional vector analysis. The integral over a closed hypersurface can be transformed into an integral over the four-volume contained within it by replacing the element of integration dS_μ by the operator

$$dS_\mu \to d\Omega\frac{\partial}{\partial x^\nu}. \qquad \text{(A2-46)}$$

For example, for the integral of a vector A^μ we have

$$\oint A^\mu dS_\mu = \int \frac{\partial A^\mu}{\partial x^\nu}d\Omega. \qquad \text{(A2-47)}$$

This formula is the generalization of Gauss' theorem. Thus, when $\partial A^\nu/\partial x^\nu = 0$, the result of integration is a true scalar and is independent of the choice of the three-surface.

B.6.3 Four-Velocity and Four-Acceleration

How do we define four vectors of velocity and acceleration? Obviously the set of the four quantities dx^μ/dt doesn't have the properties of a four-vector because dt is not an invariant. But the proper time $d\tau$ is an invariant. Observers in different frames disagree about the time interval between events, because each is using his own time axis; all agree on the value of the time interval that would be observed in the frame moving with the particle. The components of the four-velocity are therefore are defined as

$$u^\mu = \frac{dx^\mu}{d\tau}. \tag{A2-48}$$

The second equation of (A2-17) relates the proper time $d\tau$ (was dt' there) to the time dt read by a clock in frame S relative to which the object (S' frame) moves at a constant u:

$$d\tau = dt\sqrt{1 - \beta^2}.$$

We can rewrite u^μ completely in terms of quantities observed in frame S as

$$u^\mu = \frac{1}{\sqrt{1 - \beta^2}} \frac{dx^\mu}{dt} = \gamma \frac{dx^\mu}{dt}. \tag{A2-49}$$

In terms of the ordinary velocity components v_1, v_2, v_3 we have

$$u^\mu = \left(\gamma c, \gamma v_i\right), i = 1, 2, 3. \tag{A2-50}$$

The length of four-velocity must be invariant, as shown by (A2-31)

$$\sum_{\mu=0}^{3} (u^\mu)^2 = c^2. \tag{A2-51}$$

Similarly, a four-acceleration is defined as

$$w^\mu = \frac{d^2x^\mu}{d\tau^2} = \frac{du^\mu}{d\tau} \tag{A2-52}$$

Now differentiating (A2-51) with respect to τ, we obtain

$$w^\mu u^\mu = 0 \tag{A2-53}$$

thus, the four-vectors of velocity and acceleration are mutually perpendicular.

B.6.4 Four-Momentum Vector

It is obvious that Newtonian dynamics cannot hold totally. How do we know what to retain and what to discard? This is found in the generalizations that grew from the laws of motion but transcend it in their universality. These are the laws of conservation of momentum and energy. So we now generalize the definitions of momentum and energy so that in the absence of external forces the momentum and energy of a system of particles are conserved. In Newtonian mechanics the momentum $\vec{p}$ of a particle is defined as $m\vec{v}$, the product of particle's inertial mass and its velocity. A plausible generalization of this definition is to use the four-velocity u^μ and an invariant scalar m_0 that truly characterize the inertial mass of the particle and define the momentum four-vector (or four-momentum, for short) P^μ as

$$P^\mu = m_0 u^\mu. \tag{A2-54}$$

To ensure that the "mass" of the particle is truly a characteristic of the particle, it must be measured in the frame of reference in which the particle is at rest. Thus, the mass of the particle must be its proper mass. We customarily call this mass the rest mass of the particle and denote it by m_0. We can write P^μ in terms of ordinary velocity $v_i (i = 1, 2, 3)$

$$P^0 = \gamma m_0 c, \quad P^j = \gamma m_0 v_j, j = 1, 2, 3 \tag{A2-55}$$

where $\gamma = (1 - \beta^2)^{-1/2}$. We see that as $\beta = v/c \to 0$, the spatial components of the four-momentum P^μ reduce to $m_0 v_j$, the components of the ordinary momentum. This indicates that (A2-48) appears to be a reasonable generalization.

Let us write the time component P^0 as

$$P^0 = \frac{m_0 c}{\sqrt{1 - \beta^2}} = \frac{E}{c}. \tag{A2-56}$$

Now, what is the meaning of the quantity E? For low velocities, the quantity E reduces to

$$E = \frac{m_0 c^2}{\sqrt{1 - \beta^2}} \cong m_0 c^2 + \frac{1}{2} m_0 v^2.$$

The second term on the right-hand side is the ordinary kinetic energy of the particle; the first term can be interpreted as the rest energy of the particle (it is an energy the particle has, even when it is at rest), which must contain all forms of internal energy of the object, including heat energy, internal potential energy of various kinds, or rotational energy if any. Hence we can call the quantity E the total energy of the particle (moving at speed v).

We now write the four-momentum as

$$P^\mu = \left(\frac{E}{c}, P^j\right). \tag{A2-57}$$

The length of the four-momentum must be invariant, just as the length of the velocity four-vector is invariant under Lorentz transformations. We can show this easily:

$$\sum_\mu P^\mu P^\mu = \sum_\mu (m_0 u^\mu)(m_0 u^\mu) = m_0^2 c^2. \tag{A2-58}$$

But (A2-57) gives

$$\sum_\mu P^\mu P^\mu = P^2 - \frac{E^2}{c^2}.$$

Combining this with (A2-58) we arrive at the relation

$$P^2 - \frac{E^2}{c^2} = -m_0^2 c^2 \quad \text{or} \quad E^2 - P^2 c^2 = m_0^2 c^4. \tag{A2-59}$$

The total energy E and the momentum P^μ of a moving body are different when measured with respect to different reference frames. But the combination $P^2 - E^2/c^2$ has the same value for all frames of reference, namely $m_0^2 c^2$. This relation is very useful. Another very useful relation is $\vec{P} = \vec{v}(E/c^2)$. From (A2-56) we have $\gamma m_0 = E/c^2$; the second part of (A2-55) gives $\vec{P} = m_0 \vec{v}/\sqrt{1 - \beta^2}$. Combining these two yields the very useful relation $\vec{P} = \vec{v}(E/c^2)$.

The relativistic momentum, however, is not quite the familiar form found in general physics, because its spatial components contain the Lorentz factor γ. We can bring it into the old sense, and the traditional practice was to introduce a "relativistic mass" m:

$$m = m_0 \gamma = \frac{m_0}{\sqrt{1 - \beta^2}}. \tag{A2-60}$$

With this introduction of m, P^j takes the old form: $P^j = m v_j$. But some feel that the concept of relativistic mass often causes misunderstanding and vague interpretations of relativistic mechanics. So they prefer to include the factor γ, with v_j forming the proper four-velocity component u_j, and treating the mass as simply the invariant parameter m_0. For detail, see the article by Prof. Lev B Okun (The Concept of Mass, *Physics Today*, June 1989).

B.6.5 The Conservation Laws of Energy and Momentum

It is now clear that the linear momentum and energy of a particle should not be regarded as different entities, but simply as two aspects of the same attributes of the particle, since they appear as separate components of the same four-vector P^μ, that transforms according to (A2-27):

$$P^{\mu'} = L_{\nu\mu} P^\nu$$

where the Lorentz transformation matrix is given by (A2-26)

$$(L_\nu^\mu) = \begin{pmatrix} \gamma & -\beta\gamma & 0 & 0 \\ -\beta\gamma & \gamma & 0 & 0 \\ 0 & 0 & 1 & 0 \\ 0 & 0 & 0 & 1 \end{pmatrix}.$$

Thus,

$$P'^0 = \gamma (P^0 - \beta P^1),\ P'^1 = \gamma (-\beta P^0 + P^1),\ P'^2 = P^2,\ P'^3 = P^3. \quad \text{(A2-61)}$$

We see that what appears as energy in one frame appears as momentum in another frame, and vice versa.

So far, we have not discussed explicitly the conservation laws. Since linear momentum and energy are not regarded as different entities but as two aspects of the same attributes of an object, it is no longer adequate to consider linear momentum and energy separately. A natural relativistic generalization of the conservation laws of momentum and energy would be the conservation of the four-momentum. Consequently, the conservation of energy becomes one part of the law of conservation of four-momentum. This is exactly what has been found to be correct experimentally, and, in addition, this generalized conservation law of four-momentum holds for a system of particles, even when the number of particles and their rest energies are different in the initial and final states. It should be emphasized that what we mean by energy E is the total energy of an object. It consists of rest energy that contains all forms of internal energy of the body and kinetic energy. The rest energies and kinetic energies need not be individually conserved, but their sum must be. For example, in an inelastic collision, kinetic energy may be converted into some form of internal energy or vice versa, accordingly the rest energy of the object may change.

Energy and momentum conservation go together in special relativity; we cannot have one without the other. This may seem a bit puzzling for the reader, for in classical mechanics the conservation laws of energy and momentum are on different footing. That is because energy and momentum are regarded as different entities. Moreover, classical mechanics does not talk about rest energy at all.

One of the consequences of the relativistic energy-momentum generalization is the possibility of "massless" particles, which possess momentum and energy but no rest mass. From the expression for the energy and momentum of a particle

$$E = \frac{m_0 c^2}{\sqrt{1 - v^2/c^2}}, \quad \vec{P} = \frac{m_0 \vec{v}}{\sqrt{1 - v^2/c^2}} \quad \text{(A2-62)}$$

we can define a particle of zero rest mass possessing finite momentum and energy. To this purpose, we allow $v \rightarrow c$ in some inertial system S and $m_0 \rightarrow 0$ in such a way that

$$\frac{m_0}{\sqrt{1 - v^2/c^2}} = \chi \quad \text{(A2-63)}$$

remains constant. Then (A2-62) takes the simple form

$$E = \chi c^2, \quad \vec{P} = \chi c \hat{e}$$

where $\hat{e}$ is a unit vector in the direction of motion of the particle. Eliminating χ from the last two equations, we obtain

$$E = Pc, \quad \text{(A2-64)}$$

which is consistent with (A2-59): $E^2 - P^2 c^2 = m_0^2 c^4$.

Now, as $(E/c, \vec{P})$ is a four-vector, $(\chi c, \chi c \hat{e})$ is also a four-vector, the energy and momentum four-vector of a zero rest-mass particle in frame S and in any other inertial frame such as S'. It can be shown that the transformation of the energy and momentum four-vector $(\chi c, \chi c \hat{e})$ of a zero rest-mass particle is identical with that of a light wave, provided χ is made proportional to the frequency v. Thus if we associate a zero rest-mass particle with a light wave in one inertial frame, it holds in all other inertial frames. The ratio of the energy of the particle to the frequency has the dimensions of action (or angular momentum). This suggests that we can write this association by the following equations

$$E = hv \quad \text{and} \quad P = \chi c = hv/c$$

where h is Planck's constant. This massless particle of light is called a photon in modern physics, introduced by Einstein in his paper on the photoelectric effect.

B.6.6 Equivalence of Mass and Energy

The equivalence of mass and energy is the best-known relation Einstein gave in his special relativity in 1905:

$$E = mc^2 \tag{A2-65}$$

where E is the energy, m the mass, and c is the speed of light.

We can get this general idea of the equivalence of mass and energy from the consideration of electromagnetic theory. An electromagnetic field possesses energy E and momentum p, and there is a simple relationship between E and p:

$$P = E/c.$$

Thus, if an object emits light in one direction with momentum p, in order to conserve momentum, the object itself must recoil with a momentum $-p$. If we stick to the definition of momentum as $p = mv$, we may associate a "mass" with a flash of light:

$$m = \frac{p}{v} = \frac{p}{c} = \frac{E}{c^2}$$

which leads to the famous formula

$$E = mc^2.$$

This mass is not merely a mathematical fiction. Let us consider a simple thought experiment, provided by Einstein some time ago. Imagine that an emitter and absorber of light is firmly attached to the ends of a box of mass M and length L. The box is initially stationary, but is free to move. If the emitter sends a short light pulse of energy ΔE and momentum $\Delta E/c$ toward the right, the box will recoil toward the left by a small distance Δx, with momentum $p_x = -\Delta E/c$ and velocity v_x, where v_x is given by

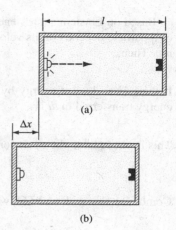

Fig. B.9 Einstein's thought experiment.

(b)

$$v_x = p_x/M = -\Delta E/cM.$$

The light pulse reaches the right end of the box approximately in time $\Delta t = L/c$ and is absorbed. The small recoil distance is then given by

$$\Delta x = v_x \Delta t = -\Delta EL/Mc^2.$$

Now, the center of mass of the system cannot move by purely internal changes and there are no external forces. It must be that the transport of energy ΔE from the left end of the box to the right end is accompanied by transport of mass Δm, so the change in the position of the center of mass of the box (denoted by δx) vanishes. The condition for this is

$$\delta x = 0 = \Delta mL + M\Delta x$$

from this we find

$$\Delta m = -\frac{M}{L}\Delta x = \frac{M}{L}\frac{\Delta EL}{Mc^2} = \Delta E/c^2,$$

or

$$\Delta E = \Delta m \cdot c^2.$$

We should not confuse the notions of equivalence and identity. The energy and mass are different physical characteristics of particles; "equivalence" only established their proportionality to each other. This is similar to the relation between the gravitational mass and inertial mass of a body; the two masses are indissolubly connected with each other and proportional to each other, but are at the same time different characteristics. The equivalence of mass and energy has been beautifully verified by experiments in which matter is annihilated and converted totally into energy. For example, when an electron and a positron, each with a rest mass m_0 come together, they disintegrate and two gamma rays emerge, each with the measured energy of m_0c^2.

Based on Einstein's mass-energy relation $E = mc^2$, we can show that the mass of a particle depends on its velocity. Let a force F act on a particle of momentum mv. Then,

$$Fdt = d(mv) \tag{A2-66}$$

If there is no loss of energy by radiation due to acceleration, then the amount of energy transferred in dt is

$$dE = c^2 dm$$

This is put equal to the work done by the force F to give

$$Fvdt = c^2 dm$$

Combining this with (A2-66), we have

$$vd(mv) = c^2 dm$$

Multiply this by m:

$$vmd(mv) = c^2 m dm,$$

integrating

$$(mv)^2 = c^2 m^2 + \text{K}.$$

K is a integration constant. Now $m = m_0$ as $v \to 0$, we find $\text{K} = -c^2 m_0^2$, and

$$m^2 v^2 = c^2(m^2 - m_0^2).$$

The m_0 is known as the rest mass of the particle. Solving for m we obtain (A2-60)

$$m = \frac{m_0}{\sqrt{1 - (v/c)^2}}.$$

It is now easy to see that a material body cannot have a velocity greater than the velocity of light. If we try to accelerate the body, as its velocity approaches the velocity of light its mass becomes larger and larger as it becomes more difficult to accelerate it further. In fact, since the mass m becomes infinite when $v = c$, we can never accelerate the body up to the speed of light.

As mentioned earlier, however, in the language of relativity theory and high-energy physics there is a trend to treat the mass as simply the invariant parameter m_0.

B.7 Problems

B.1. Observer O notes that two events are separated in space and time by $600\,m$ and 8×10^{-7} s. How fast must Observer O' be moving relative to O in order that the events be simultaneous to O'?

B.2. A meterstick makes an angle of 30° with respect to x'-axis of O'. What must be the value of v if the meterstick makes an angle of 45° with respect to the x-axis of O?

B.3. Find the speed of a particle that has a kinetic energy equal to exactly twice its rest mass energy.

B.4. Find the law of transformation of the components of a symmetric four-tensor $T^{\mu\nu}$ under Lorentz transformations.

B.5. A man on a station platform sees two trains approaching each other at the rate $7/5\ c$, but the observer on one of the trains sees the other train approaching him with a velocity $35/37\ c$. What are the velocities of the trains with respect to the station?

B.6. The equation for a spherical pulse of light starting from the origin at $t = t' = 0$ is

$$c^2 dt^2 - x^2 - y^2 - z^2 = 0.$$

Show from the Lorentz transformations that O' will observe this same pulse as spherical, in accordance with Einstein's postulate stating that the velocity of light is the same for all observers.

B.7. Referring to Figure B.4, frame S' moves with a velocity u relative to frame S along the x axis. A pair of oppositely charged plates is at rest in S' frame in a direction parallel to x' axis, and the field E between the plates is perpendicular to the plates (i.e., $\perp$ to the x' axis) and has a value that depends on the charge density σ on the plates: $E = \sigma/\varepsilon_0$. Show that view from frame S in which the plates are now moving in the x direction with a velocity u, the field E' is given by

$$E' = (1 - u^2/c^2)^{-1/2} E.$$

B.8. Referring to the previous problem, now the plates are at rest in frame S' along the y' axis. Find the electric field in both frames.

B.9. A large metallic plate moves at a constant velocity $\vec{v}$ perpendicular to a uniform magnetic field $\vec{B}$. Find the surface charge density induced on the surface of the plate.

B.10. A point charge q moves at constant velocity $\vec{v}$. Using the transformation formulas, find the magnetic field of this charge at a point whose radius vector is $\vec{r}$.

References

Bohm D (1965) *The Special Theory of Relativity*. (W. A. Benjamin, Inc., New York)
Chow TL (1995) *Classical Mechanics*. (John Wiley & Sons, New York)
Okun LB (1989) The concept of mass. Phys Today, June
Resnick R, Halliday D (1992) *Basic Concepts in Relativity*. (Macmillan Publishing Co., New York)
Terrell J (1959) Phys Rev 116:1041
Terrell J (1960) Am J Phys 28:607

Index